Nanomaterials in Industrial Chemistry

Editors

Muhammad Faheem

China University of Geosciences
Wuhan, China
and
Bahauddin Zakariya University
Multan, Pakistan

Allah Ditta

Shaheed Benazir Bhutto University Sheringal
Department of Environmental Science
Sheringal, Khyber Pakhtunkhwa
Pakistan

Jiangkun Du

School of Environment
China University of Geosciences
Wuhan, China

CRC Press
Taylor & Francis Group
Boca Raton London New York

CRC Press is an imprint of the
Taylor & Francis Group, an **informa** business

A SCIENCE PUBLISHERS BOOK

First edition published 2024
by CRC Press
2385 NW Executive Center Drive, Suite 320, Boca Raton FL 33431

and by CRC Press
4 Park Square, Milton Park, Abingdon, Oxon, OX14 4RN

CRC Press is an imprint of Taylor & Francis Group, LLC

Library of Congress Cataloging-in-Publication Data (applied for)

ISBN: 978-1-032-36952-5 (hbk)
ISBN: 978-1-032-36953-2 (pbk)
ISBN: 978-1-003-33464-4 (ebk)

DOI: 10.1201/9781003334644

Typeset in Palatino Linotype
by Prime Publishing Services

Preface

The realm of nanomaterials has swiftly become a pivotal domain within materials science and engineering, influencing everyday products with enhanced or novel properties. Nanotechnology's impact spans various scientific and technological sectors, amplifying the strengths of materials and devices while improving the efficiency of monitoring devices, environmental remediation, and renewable energy production. While the positive effects of nanotechnology are evident, it is crucial to acknowledge its negative impacts, particularly the heightened toxicological pollution resulting from the uncertain characteristics of certain nanomaterials.

Nanotechnology is inherently interdisciplinary, bridging physics, chemistry, biology, materials science, and various engineering disciplines. The scope of its applications is vast, ranging from nanomedicine, which benefits human health, to the industrial use of photocatalytic nanomaterials addressing challenges in energy, environment, sensor technology, chemical synthesis, and healthcare. Notably, nanomaterials exhibit promise in wastewater treatment, serving as catalysts, adsorbents, membranes, water disinfectants, and additives for enhanced catalytic activity.

Advancements in green synthesis methods for nanomaterials, particularly through nanobiotechnology, offer a more environmentally friendly approach. Utilizing biological systems, such as algae, bacteria, and fungi, for nanomaterial synthesis presents a superior alternative to traditional chemical and physical methods. This integration of biological processes into nanotechnology holds potential for efficient wastewater treatment with fewer environmental risks. The application of nanopharmaceuticals in designing nano-carriers for drug delivery demonstrates significant contributions to healthcare. As nanoparticle research gains recognition as a distinct discipline, it is anticipated to thrive further, with nanoparticles becoming globally accessible akin to traditional chemical reagents.

However, the rapid commercialization of nanotechnology necessitates careful research on the health and safety implications of nanomaterials.

Increased production and usage elevate the likelihood of exposure, particularly in workplaces involved in the production, processing, and disposal of nanomaterials. Nanomaterials, due to their larger surface areas, pose heightened risks to both the environment and human health compared to bulk particles. Addressing these concerns requires comprehensive risk assessments and life-cycle analyses at all stages of nanotechnology products. The selection of less toxic materials, such as graphene, can significantly mitigate environmental impacts, offering valuable insights for the education and protection of professionals and students alike in the field, including scientists, engineers, policymakers, and regulators.

Contents

CHAPTER 1

Overview of Nanomaterials
Various Nanomaterials with Different Morphologies

Muhammad Adnan Qaiser,[1,] Ahmad Hussain[2] and
Najam Ul Hassan[3]*

1. Introduction

The advanced research domain referred to as 'nanotechnology' deals with
materials of sizes at sub-nanometre or several hundred nanometre levels.
It goes without saying that over the past few decades, nanotechnology
and nanoscience have had a momentous global impact on redefining
existing research fields and creating new ones. Due to their unusual
physicochemical properties, interest in nanomaterials (NMs) has increased
recently; the discovery of nanostructures in early meteorites sparked much
interest in nanomaterials. It all started in 1857 when Michael Faraday
first reported the synthesis of gold nanoparticles (NPs). In 1959, after a
long wait, American physicist Richard Feynman addressed the American
Physical Society meeting in Caltech, saying, "There is plenty of room at the
bottom," thus inspiring the development of nanotechnology (Feynman
1992). A Japanese scientist, Norio Taniguchi, at the Tokyo University of
Science, used the term 'nanotechnology' at a conference held in 1974 to
explain semiconductor thin-film deposition and ion beam milling that

[1] School of Materials Science and Engineering, Jiangsu University, Zhenjiang, Jiangsu
212013, PR China.
[2] Department of Physics, The University of Lahore, Sargodha Campus, 40100 Sargodha,
Pakistan.
[3] Department of Physics, Division of Science and Technology, University of Education,
Lahore, 54000, Pakistan.
* Corresponding author: m.adnan.qaiser@gmail.com

exhibited a characteristic control on the nanometre scale (Sandhu 2006). He said that the separation and deformation of materials by one atom or molecule are the components of 'nanotechnology'. In 1976, Granqvist and Buhrman reported an inert gas evaporation method which became the first method for producing nanocrystals. In 1981, Eric Drexler used the term 'nanotechnology' in his first research article on nanotechnology. Eric Drexler's vision is frequently referred to as 'molecular manufacturing' or 'molecular nanotechnology' (Drexler 2004, Khan et al. 2019). Cluster science and the scanning tunnelling microscope, two important innovations gave the field of nanotechnology a boost in the early 1980s. The discovery of fullerenes in 1985 and the subsequent synthesis of 'carbon nanotubes' were the results of these advancements. In 2011, the European Commission put forth the following definition of 'nanomaterial' (Rauscher et al. 2013).

> 'Nanomaterial' means a natural, incidental or manufactured material containing particles, in an unbound state or as an aggregate or as an agglomerate and where, for 50% or more of the particles in the number size distribution, one or more external dimensions is in the size range 1–100 nm. In specific cases and where warranted by concerns for the environment, health, safety or competitiveness the number size distribution threshold of 50% may be replaced by a threshold between 1 and 50%. By derogation from the above statement, fullerenes, graphene flakes, and single wall carbon nanotubes with one or more external dimensions below 1 nm should be considered as nanomaterials.

According to this, Nanomaterials are chemical substances or materials with at least one dimension less than ~ 100 nanometre (nm or 10^{-9} m) and bulk materials are particles with all dimensions larger than 100 nm. After that, in 2012, the International Organization for Standardization ISO defined the terms associated with nanomaterials; a few of them are provided in Table 1 (Specification and Preview 2000).

2. Classification

On the basis of various criteria, NMs can be classified into various groups. Figure 1 depicts how NMs are typically categorized according to their dimensionality, morphology and chemical composition (Gleiter 2000).

2.1 Dimensionality

Based on their dimensionality and the overall shape of these materials, NMs can be further divided into four classes.

Table 1: Definition of terms associated with nanomaterials, according to ISO standards.

Terms	Definition
nanotechnology	Nanotechnology is a term used to describe science at the nanoscale level in which materials, devices, or systems are generated by manipulating shape and size at the nanoscale in order to enhance the material's special properties.
nanomaterial	A material is considered to be a nanomaterial if at least one of its dimensions falls within the range of 1–100 nm.
nanoscale	A scale with a range of 1–100 nm.
nano-object	A discrete piece of material known as a nano-object has any of its external dimensions that fall within the order of nanometres.
nanoparticle	When all of an object or particle's dimensions fall within the nanoscale range, it is referred to as a nanoparticle.
aspect ratio	The ratio of a nano-object's major axis' length to minor axis' width is known as the aspect ratio.
nanosphere	A nanoparticle with an aspect ratio of 1:21 is called a nanosphere.
nanorod	When the lengths of the longest and shortest axes differ, the term 'nanorod' is used. Nanorods have an aspect ratio greater than one, and a width between one and one hundred nanometres.
nanofiber	A nanomaterial of fiber shape with a diameter in the order of nanometres.
nanotube	Nanotubes are hollow nanofibers.

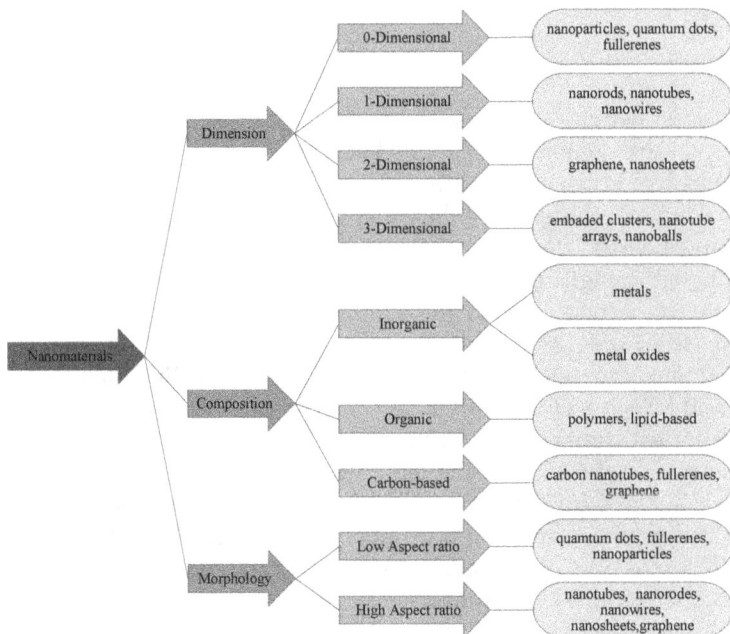

Figure 1: Classification of Nanomaterials.

1. Zero-dimensional nanomaterials (0-D): All three dimensions of this class of nanomaterials are in the nanoscale range. Fullerenes, nanoparticles and quantum dots are examples of 0-D NMs.

2. One-dimensional nanomaterials (1-D): The nanomaterials in this class have one dimension outside the nanoscale. Examples of 1-D NMs include nanotubes, nanofibers, nanorods, nanowires and nanohorns.

3. Two-dimensional nanomaterials (2-D): The nanomaterials in this class have two dimensions outside the nanoscale. Examples under this category include nanosheets, nanofilms and nanolayers.

4. Three-dimensional nanomaterials (3-D) or bulk nanomaterials: In this class, the materials are not confined to the nanoscale in any dimension. This class comprises bulk powders, dispersions of nanoparticles, arrays of nanowires and nanotubes, etc.

2.2 Chemical Composition

Nanoparticles can be made of a single material and can be compact or hollow, depending on their composition. Moreover, multiple materials can also be combined, encapsulated, coated or barcoded to form nanomaterials. NMs are often categorised into three groups according on their chemical makeup: organic, carbon-based and inorganic.

2.3 Morphological

Flatness, sphericity and aspect ratio are morphological properties. Spherical, oval, cubic, prismatic, helical and pillar morphologies are examples of low-aspect ratio shapes (Figure 2). High aspect ratio nanoparticles include nanotubes and nanowires in a variety of shapes, including helices, zigzags, belts or maybe nanowires having diameters varying with length (Figure 3). Powders, suspensions and colloids are examples of particle collections.

We shall discuss numerous nanomaterials in this chapter, based on their morphologies. Nanomaterials can be broadly categorised into high- and low-aspect ratio NMs. The ratio of an object's longest dimension to its shortest dimension is known as its aspect ratio (L/D). The aspect ratio is simply the ratio of the length to the diameter (L/D) for rod-like particles such as carbon nanotubes and nanowires; however, for planar fillers like graphene and graphite flakes, it is assumed to be the ratio of the lateral dimension to the sheet thickness. Due to more effective network development and fewer contacts needed to form a spanning cluster, fillers with higher aspect ratios percolate at lower loadings.

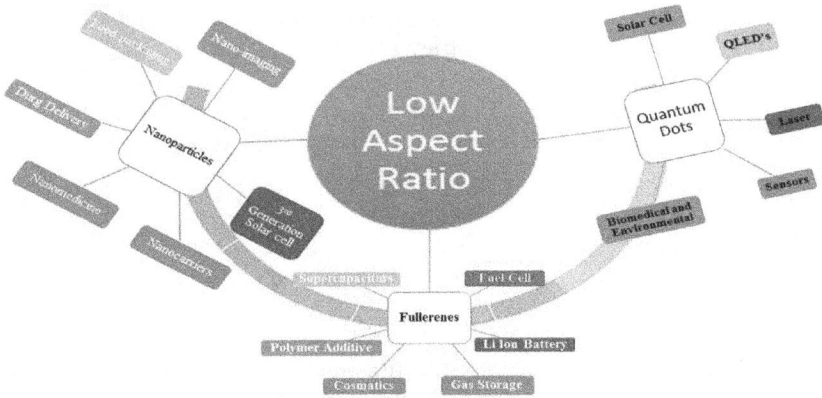

Figure 2: Applications of low-aspect ratio nanomaterials.

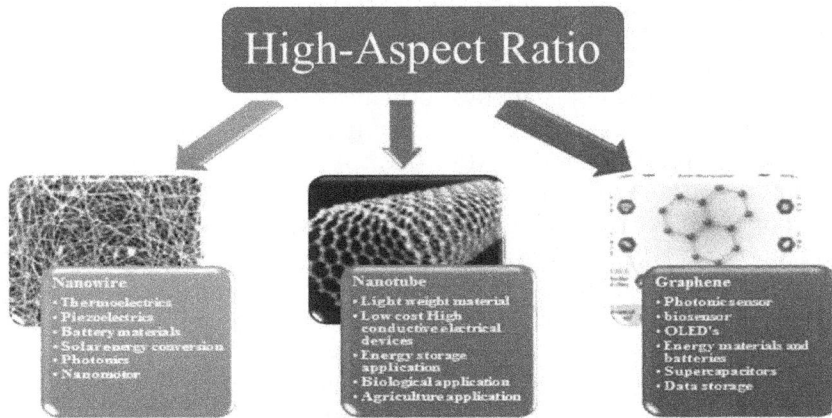

Figure 3: Applications of high-aspect ratio nanomaterials.

3. Low-Aspect Ratio Nanomaterials

3.1 Nanoparticles (NPs)

According to The International Standardization Organization (ISO), nanoparticles are defined as "Nano-objects having all exterior dimensions at the nanoscale and with minimal length differences between their longest and shortest axes." The terms 'nanofibers' or 'nanoplates' may be preferred over 'nanoparticles' if the dimensions differ noticeably (typically by more than three times). The shape, size and structure of NPs can vary–they can be spherical, cylindrical, conical, a hollow core, or spiral, among other shapes; they can also be asymmetrical (Anu Mary Ealia and Saravanakumar 2017, Rauscher et al. 2013). NPs can range in size from one to one hundred nm. The term 'atom clusters' is typically used when the size of NPs falls below 1 nm. NPs can be amorphous or crystalline with single or multiple crystal solids; they can also be agglomerated or loose (Machado et al. 2015).

NPs typically fall into one of three categories based on their composition: carbon-based, organic and inorganic (Anu Mary Ealia and Saravanakumar 2017, Rauscher et al. 2013).

3.1.1 Carbon-based NPs

NPs that are composed exclusively of carbon atoms are included in this category (Anu Mary Ealia and Saravanakumar 2017). Fullerenes, carbon black nanoparticles and carbon quantum dots are well-known examples of this group. Fullerenes are carbon molecules with a closed-cage structure that is symmetrical. C_{60} fullerenes are composed of soccer ball-shaped 60 carbon atoms (Long et al. 2013). The aggregates of highly fused spherical carbon black NPs resemble grapes (Yuan et al. 2019). Carbon nanoparticles (NPs) with sizes less than 10 nm make up carbon quantum dots (Lu et al. 2016). The unusual physicochemical properties at the nanoscale and the distinct characteristics of sp^2-hybridized carbon bonds are combined in carbon-based NPs. Because of their special electrical conductivity, high strength, electron proclivity, and their optical, warm and sorption properties (Khan et al. 2019, Mauter and Elimelech 2008), carbon-based NPs are used in a large number of applications like medication conveyance (Oh et al. 2010), energy storage (Liu et al. 2018), bioimaging (Chandra et al. 2011), photovoltaic appliances and ecological detecting applications to screen microbial nature or to distinguish microbial microorganisms (Mauter and Elimelech 2008). Carbon-based nanoparticles like nanodiamonds and carbon nano-onions are more complex. They are used in tissue engineering and drug delivery because of their biocompatibility and low toxicity (Ahlawat et al. 2021, Mochalin et al. 2012).

3.1.2 Organic NPs

NPs made of proteins, carbohydrates, lipids, polymers or any other organic compound fall under this category (Pan and Zhong 2016) organic nanoparticles (ONPs. Dendrimers, liposomes, micelles and protein complexes like ferritin are the most well-known examples in this group. These NPs are typically non-toxic and biodegradable, and may occasionally have a hollow core, such as in liposomes. Thermal and electromagnetic radiation, such as light and heat, have a negative impact on organic NPs (Anu Mary Ealia and Saravanakumar 2017). Furthermore, they are often the result of non-covalent intermolecular interactions, making them naturally more labile and providing a means of elimination from the body (Ng and Zheng 2015). The potential application area of organic NPs is determined by a variety of parameters such as composition, surface morphology, stability, carrying capacity and so on. Biomedical applications of organic NPs include cancer therapy and targeted drug delivery (Gujrati et al. 2014).

3.1.3 Inorganic NPs

NPs that are not made of carbon or organic materials belong to this category. The commonplace instances of this class are semiconductor NPs, metal NPs and ceramic NPs. Purely composed of metal precursors, metal nanoparticles (metal NPs) can be monometallic, bimetallic (Toshima and Yonezawa 1998) or polymetallic (Nascimento et al. 2018). Bimetallic NPs can be created from alloys or in various layer configurations (e.g., core-shell) (Toshima and Yonezawa 1998). These NPs have special optical and electrical properties because of the characteristics of the localised surface plasmon resonance (Khan et al. 2019). Additionally, some metal NPs have special biological, magnetic and thermal characteristics (Anu Mary Ealia and Saravanakumar 2017). Due to their increased importance as building blocks for nano-devices with a wide range of physical, chemical, biological, biomedical and pharmaceutical uses (Mody et al. 2010, Shnoudeh et al. 2019), inorganic NPs are further divided into two categories.

3.1.4 Metal-based Nanoparticles

Metals like Silver (Ag), Aluminium (Al), Copper (Cu), Gold (Au), Cadmium (Cd), Zinc (Zn), Iron (Fe), Cobalt (Co) and Lead (Pb) are all good sources of metal-based nanoparticles, Ag, Au, Cu, Fe and Zn being the most utilised metals. As a result of partially filled d-orbitals, transition metals are more redox active, which makes them the best candidates for the synthesis of metal-based NPs (Sánchez-López et al. 2020). Nanoparticle aggregation is facilitated as a result of this. The size of metal-based nanoparticles ranges from 10 to 100 nm. They come in a variety of shapes, including spherical and cylindrical ones (Toshima and Yonezawa 1998). They have peculiar properties like high surface area to volume ratios, pore size, surface charge and surface charge density, crystalline and amorphous structures, high reactivity, and sensitivity to air, moisture, heat, sunlight and other environmental factors. They have potential uses in a wide range of research fields due to their unusual properties (Mody et al. 2010, Shnoudeh et al. 2019).

3.1.5 Metal Oxide-based Nanoparticles

Metal oxide-based NPs are the oxide that have been converted from their corresponding metal-based NPs. Compared to their metal counterparts, metal oxide-based NPs have exceptional properties. Iron oxide (Fe_2O_3), Magnetite (Fe_3O_4), Aluminium oxide ($A_{12}O_3$), Cerium oxide (CeO_2), Silicon dioxide (SiO_2), Titanium oxide (TiO_2) and Zinc oxide (ZnO) are a few examples of metal oxide-based NPs (Sathyanarayanan et al. 2013). These NPs based on metal oxide have been found to be more responsive and effective. They can be found in amorphous, polycrystalline, dense, porous or hollow forms, and are typically synthesised through heating and subsequent cooling (Khan et al. 2019). Due to their high stability and high load capacity, they are primarily used in biomedical applications (Moreno-Vega et al. 2012). However, they are also used in photonics, optoelectronics, catalysis, dye degradation and other fields (D'Amato et al. 2013).

3.2 Quantum Dots

Quantum dots (QDs), also known as 'artificial atoms', are man-made nanoscale crystals that can transport electrons and were first observed by a Russian physicist Alexei Ekimov from the State Optics Institute Vavilov towards the end of the 1970s (Ekimov A. I. 1982). After that, in 1982, another Russian physicist named Alexander Efros published the first theory that sought to explain the behaviour of these very small crystals, through the confinement of their electrons (Pokutnyi 2003). American chemist Louis Brus from Bell Labs based in Murray Hill, New Jersey, attempted to create a colloidal suspension using nanocrystals, but in a liquid form, inspired by Alexei Ekimov. He published his findings in 1983 (Brus 1984) and produced the first colloidal quantum dots of cadmium sulphide that were easier to handle in this manner.

Quantum size effects can be seen in the optical and electrical characteristics of these semiconductor nanocrystals, which have nanometer-sized diameters (Ashoori 1996). Today, several QD material systems are frequently accomplished–tunable and efficient photoluminescence (PL) in particular, with narrow emission and photochemical stability, and core-shell topologies (Kubendhiran et al. 2019). The narrow emission spectra, great photochemical stability and continuous absorption spectra of QDs are only a few of their distinctive electrical and fluorescent properties. Because of their extraordinary electrical and optical properties, as well as their small size and large surface area, they are more suitable for a wide range of devices and applications. QD-based displays are just one example of the many of these applications that are now commercially available and incorporated into our day-to-day lives (Shirasaki et al. 2013). QD synthesis, characterization and applications remain a highly active area of research. The initial focus of QD research was primarily on group IV and III–V compounds; however, at the moment, QDs are based on transition-metal dichalcogenides, carbon and I–III–VI compounds, among others (Alferov 1998, Kubendhiran et al. 2019, Kuzyniak et al. 2014, Valizadeh et al. 2012). The exquisite optical properties of QDs and their role in light emission, conversion and detection are the primary foundations of their applications. As a result, QDs can be used in a lot of different ways such as:

Display applications: CsPbX3 QDs fabricated by a novel solid-state ligand-exchange method can be used in thin-film LEDs. CuInS2/ZnS QDs with stable thiol- and amine-based ligands have also been proposed for use in the production of effective film-type display devices. CdSe/CdS and CdSe/ZnS QD films coupled to Au nanorod dimer arrays can be used to alter the PL polarization of luminescent films, and polymer encapsulation of CdSe nanoplatelets and PbS QDs shows potential for applications as a composite for LEDs (Choi et al. 2019, Pachidis et al. 2019, Shiman et al. 2019, Suh et al. 2018).

Photovoltaics: In order to boost the energy conversion efficiency, photovoltaic devices have been known to incorporate CdS QDs and CdSe and CdSe/CdS core–shell QDs (Chen et al. 2013, Gudjonsdottir et al. 2019, Kokal et al. 2019). DFT calculations and theoretical research has also contributed to the creation of CdSe QD-based materials for applications in energy conversion (Cui et al. 2018).

Photodetectors: QDs have been effectively integrated with photon detecting systems based on spectral range-dependent materials to increase their performance (Konstantatos and Sargent 2010). A WS_2 QD–graphene nanocomposite has been used to create ultraviolet photodetectors, while $CsPbBr_3$ colloidal QDs and few-layer MoS_2 have been used to create phototransistors with balanced photodetection (Lin et al. 2019, Singh et al. 2019).

Biomedical and Environmental Applications: QDs are an ideal candidate for applications in bioimaging, diagnostics and biosensing, most of which are related to the biomedical and environmental sciences, due to their high luminescence, narrow emission and, depending on the composition of the elements, low toxicity and good biocompatibility (Valizadeh et al. 2012). Bioimaging and biosensing have been demonstrated by QD complexes composed of zinc chalcogenide QDs, such as undoped and Mn^{2+}-doped ZnS and surface zinc quinolate (ZnQ_2) complexes (Bhandari et al. 2019). Other examples include fibrous phosphorus QDs as fluorescent labels for human adenocarcinoma bioimaging (Amaral et al. 2020), and $CuInS_2$/ZnS QDs, MoS_2 QDs, ultrabright graphene QDs and N–S-doped graphene

QDs as markers for cell imaging (Mallick et al. 2019, Marqus et al. 2018, Schroer et al. 2019).

Catalysis: QDs have also been discovered to improve existing semiconductor photocatalysis technology and even create new catalytic routes. Carbon QD-modified graphitic carbon nitride provides photocatalytic H_2 evolution with little or no decay in activity for about half a year (Li and Zhu 2018).

3.3 Fullerenes

One crucial area of research in the existing material nanoscience is that of carbon-based materials, fullerenes taking the top spot among them. Fullerenes are zero-dimensional nanomaterials formed from vastly symmetrical hollow spheres of sp^2-hybridized carbon atoms. They are discrete molecules with an explicit number of carbon atoms, distinguishing them from other carbon allotropes. The size of fullerenes varies based on the number of carbon atoms in them, such as C_{60}, C_{70}, C_{72}, C_{76}, C_{84} and C_{100}. C_{60}, the smallest fullerene molecule (C_{60}), is composed of 60 carbon atoms bound together by covalent bonds that are sp^2-hybridised in nature. C_{60} also exhibits icosahedral symmetry and is the most often occurring fullerene in nature. Discrete peaks occur in the mass spectrum, corresponding to molecules with an exact mass of sixty or seventy or more carbon atoms, namely C_{60} and C_{70} (Kroto et al. 1985). The family of carbon-based nanomaterials was introduced to symmetry by fullerenes and they have thus opened up new possibilities in the field of nanomaterials. Consequently, new carbon-based nanostructured materials, such as carbon nanotubes and graphene, were discovered.

In 1970, a Hokkaido University researcher, Eiji Osawa, noticed that the structure of the polycyclic aromatic hydrocarbon 'corannulene' was similar to that of a soccer ball segment. The research group had proclaimed the possibility of a C_{60} shaped like a soccer ball in a Japanese journal (Yoshida and Osawa 1971). Harold Kroto et al. in 1985, discovered fullerenes in the smoky residue from carbon vaporisation in helium. In 1990, resistively heating graphite rods in an atmosphere of helium led to the development of a method for producing larger quantities of C_{60} (Krätschmer et al. 1990). Fullerenes can be found in both nature and interstellar space. They were 1991's 'molecule of the year', attracting much more research developments than any other area of study at the time.

Fullerenes have several distinguishing characteristics that make them appealing for use in healthcare, electronics, bioengineering and energy generation. Fullerenes are somewhat soluble in a variety of solvents and these properties distinguish them from other carbon allotropes (Giacalone and Martín 2010). They were the first objects for which the synergetic nature of modern science was most clearly revealed. Because C_{60} is rapidly consumed by tissues, the biological behaviour of fullerene derivatives illustrates their potential for application in the medical field. The most notable achievements for fullerenes have been obtained at the interface between chemistry and biology, and chemistry and physics, addressing both the fundamental science behind fullerenes, and their applications (Chun-Mao Lin and Tan-Yi Lu 2012). The chemical modification of fullerenes, which improves their application effectiveness, is an attractive prospect. Fullerenes can be modified in two ways (Rodríguez-Fortea et al. 2011): inner-space modification (endohedral) and outer-surface modification (exohedral).

When forming endohedral fullerenes, fullerenes are hollow cages with a rigorous nano-container for host target species (Popov et al. 2013). Due to their useful potential applications, fullerene nanocages have recently received a lot of attention in the field of materials chemistry (Mercado et al. 2008). In free space, single neutral as well as charged atoms are extremely reactive and unstable. These reactive species can be controlled in the confined environment of fullerenes; for example, the LaC_{60}^{+} ion does not react with NH_3, O_2, H_2 or NO. Reactive metals can therefore be shielded from the environment by wrapping them in fullerene cages. Endohedral fullerenes based on lithium (Li@C60) have the potential to pave the way for nanoscale lithium batteries (Bai et al. 2019, Okada et al. 2012). Gases can be stored in fullerene nanocages (Okada et al. 2012); hence, fullerene is also being investigated as a hydrogen storage material (Gaboardi et al. 2017).

Due to the ease with which their outer surfaces can be altered or functionalized, exohedral fullerenes have a wider range of potential applications. Exohedral metal doping of fullerenes has a significant impact on their electronic properties due to the transfer of electrons from the metal to the fullerene nanocage (Shokuhi Rad and Ayub 2017). As fullerene chemistry has progressed, a diverse range of functionalized fullerenes can be used in practical applications via straightforward synthetic routes (Guldi and Martin 2011). The modulation of 1D, 2D and 3D fullerene-based architectures has been made possible by combining fullerenes and hydrogen-bonding motifs (Sánchez et al. 2005). Fullerenes have demonstrated significant potential for excluding reactive oxygen species due to their excellent electron affinities; C_{60}-PDA-GSH nanoparticles have showed great promise for extracting reactive oxygen species (Zhang et al. 2019). Fullerenes have ample scope for use in drug delivery after surface modification, albeit there has been little research into their drug delivery applications (Alipour et al. 2020). Nano-vesicles based on fullerenes have been developed by researchers to delay drug release (Lin et al. 2017). Water-solvent proteins have great expectations in the field of nanomedicine. Tetra(piperazino)[60] fullerene epoxide (TPFE), a water-soluble cationic fullerene, has been used to deliver DNA and siRNA selectively to the lungs. (Minami et al. 2018).

Owing to their excellent electrochemical activity and stability during electrochemical reactions, fullerenes are regarded as suitable support materials (Fan et al. 2020, Tuktarov et al. 2020). Fullerenes have a significant impact on the stability, electronic conductivity, mass transport properties and electroactive surface area of the supported catalyst in fuel cells (Coro et al. 2016, Yousfi-Steiner et al. 2009). It can replace conventional carbon as a support material for catalysts due to their high stability and good conductivity. Additionally, fullerenes are used in the creation of efficient solar cells (Jeon et al. 2019). Fullerenes have the potential to be employed in the development of superconductors (Takabayashi and Prassides 2016). Fullerenes are useful for enhancing the mechanical properties of composites due to their strong covalent bonds as well (Rafiee et al. 2011). Furthermore, fullerenes and polymers can be combined to produce excellent thermal and flame-retardant properties (Pan et al. 2020). Advanced lubricants are created using fullerenes and their derivatives; greases and individual solid lubricants are modified using fullerenes (Tuktarov et al. 2020). Due to their anticancer, antioxidant, antibacterial and antiviral properties, fullerenes are extremely useful in medicine (Castro et al. 2017). However, fullerenes have seen their exploration slow down as other carbon-based nanomaterials were discovered. Fullerenes have the potential to boost performance due to their highly symmetrical nature and unique properties; however, more research is required for their practical application (Goodarzi et al. 2017).

4. High-Aspect Ratio Nanomaterials

4.1 Nanowires

NMs belonging to a brand-new category of semiconductors known as semiconductor nanowires have lengths ranging from hundreds of nanometres to millimetres and typical cross-sectional dimensions that can be adjusted anywhere from 1 to 100 nm (Garnett et al. 2019). The ratio of the length to the diameter (L/D) is defined as aspect ratio, which is very high in this case. Nanowire research today is heavily reliant on the technique known as the 'vapour-liquid-solid process' which was created by Dr. R. S. Wagner. Wagner noted that one of the catalysts that can be used in this vapor-liquid-solid process is gold, when he discussed the growth of silicon microwires in 1964 (Wagner and Ellis 1964). Gold is frequently used today to develop silicon nanowires, along with nickel, titanium, copper and aluminium. A high temperature in situ transmission electron microscope was used to make the very first direct *in situ* observation of nanowire growth through the above process back in 2001 (Wu and Yang 2001). This nanowire research direction quickly developed into a large, dynamic and interdisciplinary frontier over the next two decades, involving researchers from chemistry, physics, materials science and electrical engineering, among others. The rational synthesis of a large number of binary and ternary group III–V and II–VI nanowires, the growth of nanowires with controllable doping, and the synthesis of nanowires at the molecular scale are all examples of recent developments in this field. Owing to their novel electronic and photonic properties, numerous nanoscale axial and radial nanowire heterostructures have also been designed and investigated. Within the last two decades, a plethora of nanowire-based electronic and photonic, biomedical and energy conversion and storage devices have emerged as a result of the rapid development in the field of nanowires (Gudiksen et al. 2002, Lauhon et al. 2002, Wu et al. 2002).

Additionally, nanowires can be utilized for direct electrical and optical stimulation of mammalian cells (Kim et al. 2007). With their pioneering work on using nanowire transistors to probe the action potential of a single neuron cell with nanometre resolution, the Lieber group, for instance, has pushed the boundaries of what is possible between cells and nanowire devices (Patolsky et al. 2006). They have continued to be pioneers in this exciting field by recording electrical activity with unprecedented spatial and temporal resolution from fresh brain tissues and cultured cardiac cells. The field of nanowire photonics also experienced significant growth in the early 2000s, just like nanowire electronics (Huang et al. 2001). Since the first room-temperature UV nanowire laser was discovered, (Huang et al. 2001), numerous new optical processes, such as subwavelength waveguiding and nonlinear optical mixing, have been investigated for these semiconductor nanowire building blocks (Eaton et al. 2016). These investigations established the groundwork for nanowire-based subwavelength photonic integration, new nanowire scanning probe imaging and spectroscopy, and solar energy conversion. It has now opened up a brand-new, fascinating field of basic study called nanowire photonics (Yan et al. 2009).

The application of semiconductor nanowires in the conversion of solar energy, particularly in nanowire-based photovoltaics, was driven by the development of nanowire photonics. Solar energy harvesting required the introduction of a variety of nanowire heterostructure designs (Tang et al. 2011). By optimizing light absorption/trapping, charge separation and charge collection, this idea of using nanowires for photovoltaic applications is a solar cell design model system that ought to have a significant impact on the field of renewable energy. Nanowires have contributed significantly to the creation of thermoelectrics, piezoelectrics, battery materials, and solar energy conversion and storage (Hochbaum and Yang 2010). It is well known that materials with limited dimensions may exhibit distinct strain/stress responses and ion diffusion kinetics.

Since they are able to maintain electron transport along the long axis and have confinement effects across the diameter, nanowires have a lot of potential for energy storage applications.

The Yang group unveiled the first fully integrated nanowire-based direct solar water splitting system in early 2013 (Liu et al. 2013). A synthetic 'leaf' made by the same group in 2015 was a hybrid system of semiconducting nanowires and the bacteria S. Ovata (Liu et al. 2015). The bacteria use carbon dioxide and water to complete the photosynthetic process and produce a specific carbon-based butanol-like chemical while the nanowires gathered sunlight. This was the first time a fully integrated system was constructed to produce chemicals with added value solely from CO_2, water and sunlight.

Numerous nanowire applications in electronic and photonic, biomedical, and energy conversion and storage devices have already emerged as a result of the above. Each of these applications has been developed based on years of research into the synthesis, characterization, manipulation and assembly of nanowires. The entire materials, chemistry, physics and engineering research community has benefited from the path from fundamental understanding to practical applications. The future of scientific discovery, in which scientific boundaries are no longer relevant, is represented by the interdisciplinary nature of the research presented in this thematic issue (Dasgupta et al. 2014).

4.2 Nanotubes

A hollow-structured nanoscale material with two nanoscale dimensions and an aspect ratio of more than 3:1 is referred to as a nanotube (Baig et al. 2021). Sumio Iijima of the NEC Corporation in Japan, an expert in electron microscopy, discovered nanotubes in 1991 (Iijima 1991, Iijima and Ichihashi 1993). They are a new emerging generation of nanomaterials. Due to their hollow tubular nanostructure with diameters in the nanometre range, nanotubes are a distinct class of materials. Atoms can be arranged in a single sheet or multiple layers enclosed around a hollow core to make up a nanotube. The most durable fibres currently used are nanotubes, which are 10–100 times more powerful than steel (per unit weight). Researchers are trying to exploit their extraordinary strength and flexibility, and use them for making nanotube-reinforced composites with high fracture and thermal resistance. In the construction of aircraft, gears, bearings, car parts, medical devices, sports equipment and industrial food-processing equipment, such new materials could replace conventional ceramics, alumina or even metals (Gorman 2003, Zhan et al. 2003).

Among nanotube materials, carbon nanotubes (CNTs) have become one of the principal areas of interest for various examination gatherings. Single sheets of graphite with honeycomb structures are 'wound' into very long, thin carbon nanotubes with a stable, strong and flexible structure. The surfaces of these nanotubes consist of sp²-hybridized carbon particles that are organized in hexagons (Bharati et al. 2017, Scott 2005). Following the discovery of CNTs, extensive research was conducted to investigate their properties for a variety of uses. Carbon nanotubes with 1 nm diameter and a single shell were later discovered by S. Iijima and Toshinari Ichihashi(Iijima and Ichihashi 1993). Single-walled (SWNTs), double-walled (DWNTs) and multi-walled carbon nanotubes (MWNTs) are the names given to the rolled sheets which can have one, two or many walls respectively (Khan et al. 2019, Rao et al. 2001). Carbon atoms are joined by strong covalent bonds in a seamless graphitic layer that is one atom thick in single-walled carbon nanotubes (Li et al. 2019). There are two single-walled carbon nanotubes in a double-walled carbon nanotube. One carbon nanotube is settled in one more nanotube to build a two-fold walled carbon nanotube (Shen et al. 2011). Multiple sheets of single-layer carbon atoms are rolled up in multi-walled carbon nanotubes. To put it another way, numerous carbon nanotubes with a single wall are nested within one another in this case. From various kinds of nanotubes, it tends to be inferred that the nanotubes might contain one, tens or many concentric carbon shells, and these shells are isolated from one another at a distance of ~0.34 nm (Popov 2004).

The entire weight of a nanotube is concentrated in its surface layers as it is a surface structure. The uniquely large unit surface of tubulenes, which in turn predetermines their electrochemical and adsorption properties, is the result of this characteristic (Eletskii 2004). CNTs are a promising starting material for the creation of superminiaturized chemical and biological sensors due to their unparalleled unit surface and high sensitivity of the electronic properties of nanotubes to molecules adsorbed on their surface. One of the most promising uses for CNTs in electronics is in sensor devices (Akhmadishina et al. 2013, Zhang and Zhang 2009). These sensors ought to have quick response and recovery times in addition to high selectivity and sensitivity. Carbon nanotubes (CNT) can find applications in an extraordinary number of fields like added substances to polymers and impetuses, in autoelectron emanation for cathode beams of lighting parts, level showcases, gas release tubes in media transmission organizations, retention and screening of electromagnetic waves, energy transformation, lithium battery anodes, hydrogen capacity, composite materials (fillers or coatings), nanoprobes, sensors, supercapacitors and so on (Ando 2009, Dresselhaus et al. 2001). They may be utilized in a variety of industries, including the food industry, nanomedicine, medical devices, and others, due to their solid yet flexible structure and high thermal resistance.

The remarkable properties of carbon nanotubes make them valuable nanomaterials for potential practical applications. Hypothetical and exploratory investigations of carbon nanotubes have uncovered their exceptional pliable properties. J. R. Xiao and others used a logical sub-atomic underlying mechanics model to foresee SWCNT elastic qualities of 94.5 (crisscross nanotubes) and 126.2 (rocker nanotubes) GPa (Xiao et al. 2005). In another review, the Young's modulus and normal elasticity of millimetres-long multi-walled carbon nanotubes were dissected and viewed as 34.65 GPa and 0.85 GPa, respectively (Kim et al. 2017). Nanotubes made of carbon have a high aspect ratio. Carbon nanotubes are used to improve the mechanical properties of composites because of their high tensile strength.

Hundreds of tons of carbon nanotubes are produced for various applications, making them an increasingly important industrial material (Lehman et al. 2011). Carbon nanotubes are an excellent reinforcing agent due to their high aspect ratio and high tensile strength (Sahoo et al. 2010). Carbon nanotubes are used to make biodegradable, lightweight nanocomposite foams because of their low weight (Kuang et al. 2016). Carbon nanotubes can be either metallic or semiconducting, depending on their structural parameters. The use of CNTs as a central component in the design of electronic devices like rectifying diodes (Wei et al. 2006), single-electron transistors (Bockrath et al. 2001) and field-effect transistors (Rother et al. 2017) is thought to benefit from this property. Carbon nanotubes are ideal for electron field emitter applications due to their incredible structural perfection, nanoscale size, high electrical conductivity and chemical stability (Bonard et al. 2002). CNTs are a smart and valuable candidate for use in lithium-ion batteries due to their unique mechanical and electrochemical properties (Landi et al. 2009). For active lithium-ion storage, CNTs have the full potential to be used as a free-standing electrode without a binder. When compared to conventional graphite anodes, CNT-based anodes can have reversible lithium-ion capacities exceeding 1000 mAhg^{-1}. This is a significant improvement. In short, the control and optimization of the performance of CNT-based composites is influenced by the following factors (Mallakpour and Khadem 2016): (i) the carbon nanotube volume fraction, (ii) the orientation of the CNT, (iii) the CNT network grip, (iv) the aspect ratio of CNT, and (iv) the homogeneity of the composite.

In order for the system to function effectively and efficiently in some applications, it is essential to have a high concentration of CNTs dispersed in an appropriate and stable manner in water (Chik et al. 2019). Due to their hydrophobic nature, carbon nanotubes' poor dispersion in aqueous media is one of their major drawbacks. Hydrophobicity, p–p stacking and van der Waals attraction all contribute to the formation of CNT clusters. Due to their strong interactions, the CNT clusters make dissolution or dispersion in water or even organic solvent-based systems difficult (Yuan et al. 2008). CNTs' potential applications in biomedical devices, drug delivery, cell biology and drug delivery have been limited by their difficult dispersion (Chik et al. 2019). Through appropriate functionalization, carbon nanotube applications and inherent characteristics can be further tuned. The functionalization of carbon nanotubes assists researchers in controlling the properties of carbon nanotubes; without functionalization, a few properties are not feasible (Dyke and Tour 2004).

4.3 Graphene

A single layer of sp^2-hybridized carbon atoms bound in a hexagonal lattice structure in two dimensions is called graphene. In the early 1990s, Land et al. used a scanning tunneling microscope (STM) to observe monolayer graphene structures on the surface of Pt(111); this was the first encounter with graphene (Land et al. 1992). The thickness of graphite films was successfully reduced to 30 nm by Japanese scientists, Ohashi et al. later in 1997 (Ohashi et al. 1997). Finally, Novoselov et al. developed a robust and dependable method for repeatedly peeling highly oriented pyrolytic graphite (HOPG) to create monolayer graphene in 2004 (Novoselov et al. 2004).

Graphene is the main allotrope of carbon where each carbon particle is firmly clung to its neighbours by a remarkable electronic cloud, bringing to light a few outstanding issues about quantum physical science (Hughes and Walsh 2015, Tiwari et al. 2018). Fundamental research is still lacking on graphene's electronic and quantum properties. In graphene, each carbon atom is sp^2-hybridized, forming three bonds with various neighbouring carbon atoms. The s, p_x and p_y orbitals are combined in the sp^2-hybridization. Each carbon atom in the hexagonal phase has one free electron because three distinct carbon atoms bind to each other covalently and are essentially sp^2-hybridized. The p_z orbital holds this free electron, and this p-orbital lies over the plane and structures the pi bond. Interestingly, graphene's chemical and physical properties are significantly influenced by its p_z orbital (Tiwari et al. 2018, Yang et al. 2015).

As a solitary nuclear plane of carbon alongside the one of a kind quantum corridor peculiarity, graphene can be wrapped up into other graphitic materials like fullerene, nanoribbons, nanosheets, nanoplates, carbon nanotubes and thin graphene films (Geim and Novoselov 2007). Monolayer graphene exhibits an anomalous half-integer quantum Hall effect (Zhang et al. 2005), extraordinary optical properties (Blake et al. 2007, Casiraghi et al. 2007), extremely high intrinsic strength (Lee et al. 2008), superior thermal conductivity (Balandin et al. 2008) and tremendously high charge carrier mobility (Kim et al. 2009, Morozov et al. 2008, Novoselov et al. 2004) due to the extremely high internal crystal quality and massless Dirac fermions (Meyer et al. 2007, Novoselov et al. 2005b, 2005a). Due to its distinctive Dirac cone band structure close to the Fermi level, it is a zero-gap semiconductor with an exceptionally high concentration of charge carriers and ballistic transport. The development of artificial human-made materials with tunable band gaps that can be useful for the next generation of computing opens up a number of new opportunities thanks to this property. Quantum effects in graphene can also be studied at room temperature because massless electrons can move through the honeycomb lattice without scattering over a sub-micrometre distance (Geim 2009).

The material known as graphene is thought to be revolutionary. The domains of supercapacitors, field-effect transistors, solar cells, batteries, catalysis, sensors and membrane technology have all been promised to be transformed by graphene's unique properties (Amiri et al. 2015, Anh et al. 2020, Blaschke et al. 2016, Dalapati et al. 2020, G. Liu et al. 2015, Novoselov et al. 2007). The possibilities for using graphene are truly limitless, and many of them are yet to be thought of. In order to meet the requirements of various devices and achieve desirable performance, graphene fever has continuously grown. As a result, more pressure is being placed on this research to investigate ways to market graphene.

The transistor made using graphene is a nanoscale device that works by crossing just one electron at once. Since its inception, this kind of transistor has received a lot of attention, and many of them are now available for everyday use. The fact that graphene-based transistors can function effectively at room temperature and with high sensitivity at low voltage is their primary advantage. In addition to advancing microchip technology, these characteristics make graphene-based transistors superior to silicon-based transistors (Blaschke et al. 2016, He et al. 2017).

Graphene is the most desirable material for sensors because of its unique optical properties, high thermal conductivity, excellent electrical conductivity, high carrier mobility and density, and large surface-to-volume ratio, among other properties (Cohen-Tanugi and Grossman 2014, Goli et al. 2014, Zhao et al. 2016). Several of the properties of graphene are useful in sensor applications. Biosensors, diagnostics, field-effect transistors, DNA sensors and gas sensors are just a few of the many applications for graphene as a whole or as a component of sensors (Cohen-Tanugi and Grossman 2014, Zhao et al. 2016). The standout electrical conductivity, high angle proportion and dispersibility of graphene are considerably better than the traditional inorganic-based cathode while relieving the terminals of their imperatives. Graphene has been used frequently in supercapacitors, energy components, lithium-ion batteries and Li-S batteries due to its adaptable behaviour. Li-S batteries can produce energies of up to 500 Wh/kg and even more when used in real time (Bogue 2014, Song et al. 2016).

With a field improvement factor of up to 3700, graphene can activate an electric field at 0.1 V/m^2. Graphene displays are currently available in the market and used in a variety of applications. In this contest, George et al. investigated the mechanical behavior of a model touch panel display, which comprises of two layers of CVD grown pure graphene entrenched into PET films in tension and under contact-stress dynamic loading. This study suggests that graphene might be the best material for flexible touch panel displays of the next generation (Anagnostopoulos et al. 2016, Dubey et al. 2023).

5. Conclusion and Future Prospects

Nanomaterials' exceptional tunable properties and improved performance compared to their bulk counterparts have given them prominence in technological advancement. Due to their exceptional and distinctive qualities, these nanomaterials are revolutionising numerous industrial applications. Any material with a dimension between 1 and 100 nm is considered as a nanomaterial. When compared to micromaterials or bulk materials, the surface effects of nanomaterials are distinct. For instance, nanomaterials have a lower binding energy per atom as they have fewer direct neighbours for the atoms at the surface. According to the Gibbs–Thomson equation, this change directly affects the melting temperature of nanomaterials. Because of the larger reaction surface, nanomaterials with larger surface areas and high surface-to-volume ratios typically exhibit significant effects of surface properties on their structure. One important aspect of the surface effects is the dispersion of nanomaterials. The agglomeration and aggregation of nanomaterials, which has a negative impact on both their surface area and their properties at the nanoscale, can be caused by the strong and attractive interactions that occur between particles.

In the range of 1–100 nm, where quantum phenomena are present, nanomaterials exhibit distinct size-dependent properties. The impact of quantum confinement becomes apparent as the material radius approaches the asymptotic exciton Bohr radius (the separation between the electron and the hole). In other words, nanomaterials become quantal as a result of the material's reduced size, which intensifies the quantum effects. These quantum structures are real physical structures in which all the charge carriers (electron and hole) are constrained to the real physical dimensions. Some non-magnetic bulk materials, like palladium, platinum and gold, start to become magnetic at the nanoscale due to quantum confinement effects for example. Significant changes in electron affinity or the capacity to accept or donate electrical charges can also be brought on by quantum confinement, having an immediate impact on the material's catalytic capabilities.

Nanomaterials with a carbon base, nanoporous materials, core-shell materials, ultrathin 2-dimensional nanomaterials and metal-based nanomaterials are all included in the nanomaterials family. The size, composition, shape and origin of NMs are what distinguish them from each other. The value of each classification rises when it is possible to anticipate the distinctive properties of NMs. Hence, the following morphological characteristics should be considered: aspect ratio, sphericity and flatness. For planar fillers like graphene and graphite flakes, the aspect ratio is taken to be the ratio of the lateral dimension to the sheet thickness. For other materials, the aspect ratio is simply the ratio of the length to the diameter (L/D). Nanotubes and nanowires with a high aspect ratio come in a variety of shapes, including helixes, zigzags, belts and even nanowires whose diameter varies with length. Shapes with a low aspect ratio include spherical, oval, cubic, helical and pillar shapes. Particles can be gathered into powders, suspensions or colloids. Low-aspect ratio nanomaterials have found potential uses in a variety of industries like drug delivery, energy storage, bioimaging, photovoltaic devices, microbial nature detection in the environment, photonics, optoelectronics, catalysis, dye degradation, etc. Whereas, high-aspect ratio nanomaterials are better suited to applications in thermoelectrics, piezoelectrics, battery materials, solar energy conversion and storage, gas release tubes in media transmission organisations, retention and screening of electromagnetic waves, energy transformation, lithium battery anodes, hydrogen capacity, composite materials (fillers or coatings), nanoprobes, sensors, superconductors, autoelectron emanation for cathode beams of lighting parts, level showcases, and composite materials.

Technological developments in the area of nanotechnology are related to the future of advanced technology. With the development of nanomaterial-based engineering strategies, the dream of producing clean energy is becoming a reality. These materials have demonstrated promising results, including the development of new generations of hydrogen fuel cells and solar cells, their capacity for hydrogen storage, their effectiveness as water splitting catalysts, and more. The efficiency with which materials can be manipulated on the nanoscale for various purposes determines the future technologies. Be that as it may, the turn of events and viable use of nanomaterials include many difficulties simultaneously. For instance, the presence of defects in nanomaterials can compromise their inherent properties and impact their performance. The amalgamation of nanomaterials through savvy courses is another significant test. Large-scale production of high-quality nanomaterials is typically restricted by the need for sophisticated equipment and harsh conditions. The use of nanomaterials in industry is rising and there is also a growing demand for nanoscale material production. Furthermore, the scope of nanotechnology research is vast; new nanomaterials with fascinating properties will continue to be studied, and additional fields will be discovered in the future.

References

Ahlawat, J., Masoudi Asil, S., Guillama Barroso, G., Nurunnabi, M. and Narayan, M. 2021. Application of carbon nano onions in the biomedical field: recent advances and challenges. Biomater. Sci. 9: 626–644. https://doi.org/10.1039/D0BM01476A.

Akhmadishina, K.F., Bobrinetskii, I.I., Komarov, I.A., Malovichko, A.M., Nevolin, V.K., Petukhov, V.A. et al. 2013. Flexible biological sensors based on carbon nanotube films. Nanotechnologies Russ. 8: 721–726. https://doi.org/10.1134/S1995078013060025.

Alferov, Z.I. 1998. The history and future of semiconductor heterostructures. Semiconductors 32: 1–14. https://doi.org/10.1134/1.1187350.

Alipour, E., Alimohammady, F., Yumashev, A. and Maseleno, A. 2020. Fullerene C60 containing porphyrin-like metal center as drug delivery system for ibuprofen drug. J. Mol. Model. 26: 7. https://doi.org/10.1007/s00894-019-4267-1.

Amaral, P.E.M., Hall, D.C., Pai, R., Król, J.E., Kalra, V., Ehrlich, G.D. et al. 2020. Fibrous phosphorus quantum dots for cell imaging. ACS Appl. Nano Mater. 3: 752–759. https://doi.org/10.1021/acsanm.9b01786.

Amiri, A., Sadri, R., Shanbedi, M., Ahmadi, G., Chew, B.T., Kazi, S.N. et al. 2015. Performance dependence of thermosyphon on the functionalization approaches: An experimental study on thermo-physical properties of graphene nanoplatelet-based water nanofluids. Energy Convers. Manag. 92: 322–330. https://doi.org/10.1016/j.enconman.2014.12.051.

Anagnostopoulos, G., Pappas, P.-N., Li, Z., Kinloch, I.A., Young, R.J., Novoselov, K.S. et al. 2016. Mechanical stability of flexible graphene-based displays. ACS Appl. Mater. Interfaces 8: 22605–22614. https://doi.org/10.1021/acsami.6b05227.

Ando, T. 2009. The electronic properties of graphene and carbon nanotubes. NPG Asia Mater. 1: 17–21. https://doi.org/10.1038/asiamat.2009.1.

Anh, N.N., Van Chuc, N., Thang, B.H., Van Nhat, P., Hao, N., Phuong, D.D. et al. 2020. Solar Cell Based on Hybrid Structural SiNW/Poly(3,4ethylenedioxythiophene): Poly(styrenesulfonate)/Graphene. Glob. Challenges 4: 2000010. https://doi.org/10.1002/gch2.202000010.

Anu Mary Ealia, S. and Saravanakumar, M.P. 2017. A review on the classification, characterisation, synthesis of nanoparticles and their application. IOP Conf. Ser. Mater. Sci. Eng. 263: 032019. https://doi.org/10.1088/1757-899X/263/3/032019.

Ashoori, R.C. 1996. Electrons in artificial atoms. Nature 379: 413–419. https://doi.org/10.1038/379413a0.

Bai, H., Gao, H., Feng, W., Zhao, Y. and Wu, Y. 2019. Interaction in Li@Fullerenes and Li+@ Fullerenes: First Principle Insights to Li-Based Endohedral Fullerenes. Nanomaterials 9: 630. https://doi.org/10.3390/nano9040630.

Baig, N., Kammakakam, I., Falath, W. and Kammakakam, I. 2021. Nanomaterials: A review of synthesis methods, properties, recent progress, and challenges. Mater. Adv. 2: 1821–1871. https://doi.org/10.1039/d0ma00807a.

Balandin, A.A., Ghosh, S., Bao, W., Calizo, I., Teweldebrhan, D., Miao, F. et al. 2008. Superior thermal conductivity of single-layer graphene. Nano Lett. 8: 902–907. https://doi.org/10.1021/nl0731872.

Bhandari, S., Pramanik, S., Biswas, N.K., Roy, S. and Pan, U.N. 2019. Enhanced luminescence of a quantum dot complex following interaction with protein for applications in cellular imaging, sensing, and white-light generation. ACS Appl. Nano Mater. 2: 2358–2366. https://doi.org/10.1021/acsanm.9b00233.

Bharati, R., Sundaramurthy, S. and Thakur, C. 2017. Nanomaterials and food-processing wastewater. pp. 479–516. *In*: Water Purification. Elsevier. https://doi.org/10.1016/B978-0-12-804300-4.00014-9.

Blake, P., Hill, E.W., Castro Neto, A.H., Novoselov, K.S., Jiang, D., Yang, R. et al. 2007. Making graphene visible. Appl. Phys. Lett. 91: 063124. https://doi.org/10.1063/1.2768624.

Blaschke, B.M., Lottner, M., Drieschner, S., Calia, A.B., Stoiber, K., Rousseau, L. et al. 2016. Flexible graphene transistors for recording cell action potentials. 2D Mater. 3: 025007. https://doi.org/10.1088/2053-1583/3/2/025007.

Bockrath, M., Liang, W., Bozovic, D., Hafner, J.H., Lieber, C.M., Tinkham, M. et al. 2001. Resonant electron scattering by defects in single-walled carbon nanotubes. Science. 291: 283–285. https://doi.org/10.1126/science.291.5502.283.

Bogue, R. 2014. Graphene sensors: a review of recent developments. Sens. Rev. 34: 233–238. https://doi.org/10.1108/SR-03-2014-631.

Bonard, J.-M., Croci, M., Klinke, C., Kurt, R., Noury, O. and Weiss, N. 2002. Carbon nanotube films as electron field emitters. Carbon N. Y. 40: 1715–1728. https://doi.org/10.1016/S0008-6223(02)00011-8.

Brus, L.E. 1984. Electron–electron and electron-hole interactions in small semiconductor crystallites: The size dependence of the lowest excited electronic state. J. Chem. Phys. 80: 4403–4409. https://doi.org/10.1063/1.447218.

Casiraghi, C., Hartschuh, A., Lidorikis, E., Qian, H., Harutyunyan, H., Gokus, T. et al. 2007. Rayleigh imaging of graphene and graphene layers. Nano Lett. 7: 2711–2717. https://doi.org/10.1021/nl071168m.

Castro, E., Garcia, A.H., Zavala, G. and Echegoyen, L. 2017. Fullerenes in biology and medicine. J. Mater. Chem. B 5: 6523–6535. https://doi.org/10.1039/C7TB00855D.

Chandra, S., Das, P., Bag, S., Laha, D. and Pramanik, P. 2011. Synthesis, functionalization and bioimaging applications of highly fluorescent carbon nanoparticles. Nanoscale 3: 1533. https://doi.org/10.1039/c0nr00735h.

Chen, G., Seo, J., Yang, C. and Prasad, P.N. 2013. Nanochemistry and nanomaterials for photovoltaics. Chem. Soc. Rev. 42: 8304. https://doi.org/10.1039/c3cs60054h.

Chik, M.W., Hussain, Z., Zulkefeli, M., Tripathy, M., Kumar, S., Majeed, A.B.A. et al. 2019. Polymer-wrapped single-walled carbon nanotubes: a transformation toward better applications in healthcare. Drug Deliv. Transl. Res. 9: 578–594. https://doi.org/10.1007/s13346-018-0505-9.

Choi, J., Choi, W. and Jeon, D.Y. 2019. Ligand-Exchange-Ready CuInS2/ZnS quantum dots via surface-ligand composition control for film-type display devices. ACS Appl. Nano Mater. 2: 5504–5511. https://doi.org/10.1021/acsanm.9b01085.

Chun-Mao Lin. and Tan-Yi Lu. 2012. C60 fullerene derivatized nanoparticles and their application to therapeutics. Recent Pat. Nanotechnol. 6: 105–113. https://doi.org/10.2174/187221012800270135.

Cohen-Tanugi, D. and Grossman, J.C. 2014. Mechanical strength of nanoporous graphene as a desalination membrane. Nano Lett. 14: 6171–6178. https://doi.org/10.1021/nl502399y.

Coro, J., Suárez, M., Silva, L.S.R., Eguiluz, K.I.B. and Salazar-Banda, G.R. 2016. Fullerene applications in fuel cells: A review. Int. J. Hydrogen Energy 41: 17944–17959. https://doi.org/10.1016/j.ijhydene.2016.08.043.

Cui, P., Tamukong, P.K. and Kilina, S. 2018. Effect of binding geometry on charge transfer in CdSe nanocrystals functionalized by N719 dyes to tune energy conversion efficiency. ACS Appl. Nano Mater. 1: 3174–3185. https://doi.org/10.1021/acsanm.8b00350.

D'Amato, R., Falconieri, M., Gagliardi, S., Popovici, E., Serra, E., Terranova, G. et al. 2013. Synthesis of ceramic nanoparticles by laser pyrolysis: From research to applications. J. Anal. Appl. Pyrolysis 104: 461–469. https://doi.org/10.1016/j.jaap.2013.05.026.

Dalapati, G.K., Masudy-Panah, S., Moakhar, R.S., Chakrabortty, S., Ghosh, S., Kushwaha, A. et al. 2020. Nanoengineered advanced materials for enabling hydrogen economy:

functionalized graphene–incorporated cupric oxide catalyst for efficient solar hydrogen production. Glob. Challenges 4: 1900087. https://doi.org/10.1002/gch2.201900087.

Dasgupta, N.P., Sun, J., Liu, C., Brittman, S., Andrews, S.C., Lim, J. et al. 2014. 25th Anniversary Article: Semiconductor Nanowires - Synthesis, Characterization, and Applications. Adv. Mater. 26: 2137–2184. https://doi.org/10.1002/adma.201305929.

Dresselhaus, M.S., Dresselhaus, G. and Avouris, P. (eds.). 2001. Carbon Nanotubes, Topics in Applied Physics. Springer Berlin Heidelberg, Berlin, Heidelberg. https://doi.org/10.1007/3-540-39947-X.

Drexler, K.E. 2004. Nanotechnology: From feynman to funding. Bull. Sci. Technol. Soc. 24: 21–27. https://doi.org/10.1177/0270467604263113.

Dubey, P.K., Hong, J., Lee, K. and Singh, P. 2023. Graphene-based materials: synthesis and applications. pp. 59–84. *In*: Nanomaterials. Springer Nature Singapore, Singapore. https://doi.org/10.1007/978-981-19-7963-7_3.

Dyke, C.A. and Tour, J.M. 2004. Covalent functionalization of single-walled carbon nanotubes for materials applications. J. Phys. Chem. A 108: 11151–11159. https://doi.org/10.1021/jp046274g.

Eaton, S.W., Fu, A., Wong, A.B., Ning, C.-Z. and Yang, P. 2016. Semiconductor nanowire lasers. Nat. Rev. Mater. 1: 16028. https://doi.org/10.1038/natrevmats.2016.28.

Ekimov, A.I. and Onushchenko, A.A. 1982. Quantum size effect in the optical-spectra of semiconductor micro-crystals. Sov. Phys. Semicond. 16: 775–778.

Eletskii, A.V. 2004. Sorption properties of carbon nanostructures. Physics-Uspekhi 47: 1119–1154. https://doi.org/10.1070/PU2004v047n11ABEH002017.

Fan, X., Soin, N., Li, Haitao, Li, Hua, Xia, X. and Geng, J. 2020. Fullerene (C60) Nanowires: The preparation, characterization, and potential applications. Energy Environ. Mater. 3: 469–491. https://doi.org/10.1002/eem2.12071.

Feynman, R.P. 1992. There's plenty of room at the bottom [data storage]. J. Microelectromechanical Syst. 1: 60–66. https://doi.org/10.1109/84.128057.

Gaboardi, M., Sarzi Amadé, N., Aramini, M., Milanese, C., Magnani, G., Sanna, S. et al. 2017. Extending the hydrogen storage limit in fullerene. Carbon N. Y. 120: 77–82. https://doi.org/10.1016/j.carbon.2017.05.025.

Garnett, E., Mai, L. and Yang, P. 2019. Introduction: 1D Nanomaterials/Nanowires. Chem. Rev. 119: 8955–8957. https://doi.org/10.1021/acs.chemrev.9b00423.

Geim, A.K. 2009. Graphene: Status and Prospects. Science. 324: 1530–1534. https://doi.org/10.1126/science.1158877.

Geim, A.K. and Novoselov, K.S. 2007. The rise of graphene. Nat. Mater. 6: 183–191. https://doi.org/10.1038/nmat1849.

Giacalone, F. and Martín, N. 2010. New concepts and applications in the macromolecular chemistry of fullerenes. Adv. Mater. 22: 4220–4248. https://doi.org/10.1002/adma.201000083.

Gleiter, H. 2000. Nanostructured materials: basic concepts and microstructure. Acta Mater. 48: 1–29. https://doi.org/10.1016/S1359-6454(99)00285-2.

Goli, P., Ning, H., Li, X., Lu, C.Y., Novoselov, K.S. and Balandin, A.A. 2014. Thermal properties of graphene–copper–graphene heterogeneous films. Nano Lett. 14: 1497–1503. https://doi.org/10.1021/nl404719n.

Goodarzi, S., Da Ros, T., Conde, J., Sefat, F. and Mozafari, M. 2017. Fullerene: biomedical engineers get to revisit an old friend. Mater. Today 20: 460–480. https://doi.org/10.1016/j.mattod.2017.03.017.

Gorman, J. 2003. Fracture protection. Sci. News 163: 3. https://doi.org/10.2307/4014350.

Gudiksen, M.S., Lauhon, L.J., Wang, J., Smith, D.C. and Lieber, C.M. 2002. Growth of nanowire superlattice structures for nanoscale photonics and electronics. Nature 415: 617–620. https://doi.org/10.1038/415617a.

Gudjonsdottir, S., van der Stam, W., Koopman, C., Kwakkenbos, B., Evers, W.H. and Houtepen, A.J. 2019. On the stability of permanent electrochemical doping of quantum dot, fullerene, and conductive polymer films in frozen electrolytes for use in semiconductor devices. ACS Appl. Nano Mater. 2: 4900–4909. https://doi.org/10.1021/acsanm.9b00863.

Gujrati, M., Malamas, A., Shin, T., Jin, E., Sun, Y. and Lu, Z.-R. 2014. Multifunctional cationic lipid-based nanoparticles facilitate endosomal escape and reduction-triggered cytosolic siRNA release. Mol. Pharm. 11: 2734–2744. https://doi.org/10.1021/mp400787s.

Guldi, D.M. and Martin, N. 2011. Functionalized fullerenes: synthesis and functions. pp. 379–398. *In*: Comprehensive Nanoscience and Technology. Elsevier. https://doi.org/10.1016/B978-0-12-374396-1.00007-6.

He, S., Qian, Y., Liu, K., Macosko, C.W. and Stein, A. 2017. Modified-graphene-oxide-containing styrene masterbatches for thermosets. Ind. Eng. Chem. Res. 56: 11443–11450. https://doi.org/10.1021/acs.iecr.7b02583.

Hochbaum, A.I. and Yang, P. 2010. Semiconductor nanowires for energy conversion. Chem. Rev. 110: 527–546. https://doi.org/10.1021/cr900075v.

Huang, M.H., Mao, S., Feick, H., Yan, H., Wu, Y., Kind, H. et al. 2001. Room-temperature ultraviolet nanowire nanolasers. Science. 292: 1897–1899. https://doi.org/10.1126/science.1060367.

Hughes, Z.E. and Walsh, T.R. 2015. Computational chemistry for graphene-based energy applications: progress and challenges. Nanoscale 7: 6883–6908. https://doi.org/10.1039/C5NR00690B.

Iijima, S. 1991. Helical microtubules of graphitic carbon. Nature 354: 56–58. https://doi.org/10.1038/354056a0.

Iijima, S. and Ichihashi, T. 1993. Single-shell carbon nanotubes of 1-nm diameter. Nature 363: 603–605. https://doi.org/10.1038/363603a0.

Jeon, I., Shawky, A., Lin, H.-S., Seo, S., Okada, H., Lee, J.-W. et al. 2019. Controlled redox of lithium-ion endohedral fullerene for efficient and stable metal electrode-free perovskite solar cells. J. Am. Chem. Soc. 141: 16553–16558. https://doi.org/10.1021/jacs.9b06418.

Khan, I., Saeed, K. and Khan, I. 2019. Nanoparticles: Properties, applications and toxicities. Arab. J. Chem. 12: 908–931. https://doi.org/10.1016/j.arabjc.2017.05.011.

Kim, H., Wang, M., Lee, S.K., Kang, J., Nam, J.-D., Ci, L. et al. 2017. Tensile properties of millimeter-long multi-walled carbon nanotubes. Sci. Rep. 7: 9512. https://doi.org/10.1038/s41598-017-10279-0.

Kim, Keun Soo., Zhao, Y., Jang, H., Lee, S.Y., Kim, J.M., Kim, Kwang S. et al. 2009. Large-scale pattern growth of graphene films for stretchable transparent electrodes. Nature 457: 706–710. https://doi.org/10.1038/nature07719.

Kim, W., Ng, J.K., Kunitake, M.E., Conklin, B.R. and Yang, P. 2007. Interfacing silicon nanowires with mammalian cells. J. Am. Chem. Soc. 129: 7228–7229. https://doi.org/10.1021/ja071456k.

Kokal, R.K., Bredar, A.R.C., Farnum, B.H. and Deepa, M. 2019. Solid-state succinonitrile/sulfide hole transport layer and carbon fabric counter electrode for a quantum dot solar cell. ACS Appl. Nano Mater. 2: 7880–7887. https://doi.org/10.1021/acsanm.9b01873.

Konstantatos, G. and Sargent, E.H. 2010. Nanostructured materials for photon detection. Nat. Nanotechnol. 5: 391–400. https://doi.org/10.1038/nnano.2010.78.

Krätschmer, W., Lamb, L.D., Fostiropoulos, K. and Huffman, D.R. 1990. Solid C60: a new form of carbon. Nature 347: 354–358. https://doi.org/10.1038/347354a0.

Kroto, H.W., Heath, J.R., O'Brien, S.C., Curl, R.F. and Smalley, R.E. 1985. C60: Buckminsterfullerene. Nature 318: 162–163. https://doi.org/10.1038/318162a0.

Kuang, T., Chang, L., Chen, F., Sheng, Y., Fu, D. and Peng, X. 2016. Facile preparation of lightweight high-strength biodegradable polymer/multi-walled carbon nanotubes nanocomposite foams for electromagnetic interference shielding. Carbon N. Y. 105: 305–313. https://doi.org/10.1016/j.carbon.2016.04.052.

Kubendhiran, S., Bao, Z., Dave, K. and Liu, R.-S. 2019. Microfluidic synthesis of semiconducting colloidal quantum dots and their applications. ACS Appl. Nano Mater. 2: 1773–1790. https://doi.org/10.1021/acsanm.9b00456.

Kuzyniak, W., Adegoke, O., Sekhosana, K., D'Souza, S., Tshangana, S.C., Hoffmann, B. et al. 2014. Synthesis and characterization of quantum dots designed for biomedical use. Int. J. Pharm. 466: 382–389. https://doi.org/10.1016/j.ijpharm.2014.03.037.

Land, T.A., Michely, T., Behm, R.J., Hemminger, J.C. and Comsa, G. 1992. STM investigation of single layer graphite structures produced on Pt(111) by hydrocarbon decomposition. Surf. Sci. 264: 261–270. https://doi.org/10.1016/0039-6028(92)90183-7.

Landi, B.J., Ganter, M.J., Cress, C.D., DiLeo, R.A. and Raffaelle, R.P. 2009. Carbon nanotubes for lithium ion batteries. Energy Environ. Sci. 2: 638. https://doi.org/10.1039/b904116h.

Lauhon, L.J., Gudiksen, M.S., Wang, D. and Lieber, C.M. 2002. Epitaxial core–shell and core–multishell nanowire heterostructures. Nature 420: 57–61. https://doi.org/10.1038/nature01141.

Lee, C., Wei, X., Kysar, J.W. and Hone, J. 2008. Measurement of the elastic properties and intrinsic strength of monolayer graphene. Science. 321: 385–388. https://doi.org/10.1126/science.1157996.

Lehman, J.H., Terrones, M., Mansfield, E., Hurst, K.E. and Meunier, V. 2011. Evaluating the characteristics of multiwall carbon nanotubes. Carbon N. Y. 49: 2581–2602. https://doi.org/10.1016/j.carbon.2011.03.028.

Li, L. and Zhu, X. 2018. Enhanced photocatalytic hydrogen evolution of carbon quantum dot modified 1D protonated nanorods of graphitic carbon nitride. ACS Appl. Nano Mater. 1: 5337–5344. https://doi.org/10.1021/acsanm.8b01381.

Li, X.-Q., Hou, P.-X., Liu, C. and Cheng, H.-M. 2019. Preparation of metallic single-wall carbon nanotubes. Carbon N. Y. 147: 187–198. https://doi.org/10.1016/j.carbon.2019.02.089.

Lin, M.-S., Chen, R.-T., Yu, N.-Y., Sun, L.-C., Liu, Y., Cui, C.-H. et al. 2017. Fullerene-based amino acid ester chlorides self-assembled as spherical nano-vesicles for drug delayed release. Colloids Surfaces B Biointerfaces 159: 613–619. https://doi.org/10.1016/j.colsurfb.2017.08.007.

Lin, R., Li, X., Zheng, W. and Huang, F. 2019. Balanced photodetection in mixed-dimensional phototransistors consisting of $CsPbBr_3$ quantum dots and few-layer MoS_2. ACS Appl. Nano Mater. 2: 2599–2605. https://doi.org/10.1021/acsanm.9b00558.

Liu, C., Gallagher, J.J., Sakimoto, K.K., Nichols, E.M., Chang, C.J., Chang, M.C.Y. et al. 2015. Nanowire–bacteria hybrids for unassisted solar carbon dioxide fixation to value-added chemicals. Nano Lett. 15: 3634–3639. https://doi.org/10.1021/acs.nanolett.5b01254.

Liu, C., Tang, J., Chen, H.M., Liu, B. and Yang, P. 2013. A fully integrated nanosystem of semiconductor nanowires for direct solar water splitting. Nano Lett. 13: 2989–2992. https://doi.org/10.1021/nl401615t.

Liu, G., Jin, W. and Xu, N. 2015. Graphene-based membranes. Chem. Soc. Rev. 44: 5016–5030. https://doi.org/10.1039/C4CS00423J.

Liu, M., Zhao, F., Zhu, D., Duan, H., Lv, Y., Li, L. et al. 2018. Ultramicroporous carbon nanoparticles derived from metal–organic framework nanoparticles for high-performance supercapacitors. Mater. Chem. Phys. 211: 234–241. https://doi.org/10.1016/j.matchemphys.2018.02.030.

Long, C.M., Nascarella, M.A. and Valberg, P.A. 2013. Carbon black vs. black carbon and other airborne materials containing elemental carbon: Physical and chemical distinctions. Environ. Pollut. 181: 271–286. https://doi.org/10.1016/j.envpol.2013.06.009.

Lu, K.-Q., Quan, Q., Zhang, N. and Xu, Y.-J. 2016. Multifarious roles of carbon quantum dots in heterogeneous photocatalysis. J. Energy Chem. 25: 927–935. https://doi.org/10.1016/j.jechem.2016.09.015.

Machado, S., Pacheco, J.G., Nouws, H.P.A., Albergaria, J.T. and Delerue-Matos, C. 2015. Characterization of green zero-valent iron nanoparticles produced with tree leaf extracts. Sci. Total Environ. 533: 76–81. https://doi.org/10.1016/j.scitotenv.2015.06.091.

Mallakpour, S. and Khadem, E. 2016. Carbon nanotube–metal oxide nanocomposites: Fabrication, properties and applications. Chem. Eng. J. 302: 344–367. https://doi.org/10.1016/j.cej.2016.05.038.

Mallick, S., Kumar, P. and Koner, A.L. 2019. Freeze-resistant cadmium-free quantum dots for live-cell imaging. ACS Appl. Nano Mater. 2: 661–666. https://doi.org/10.1021/acsanm.8b02231.

Marqus, S., Ahmed, H., Ahmed, M., Xu, C., Rezk, A.R. and Yeo, L.Y. 2018. Increasing exfoliation yield in the synthesis of mos_2 quantum dots for optoelectronic and other applications through a continuous multicycle acoustomicrofluidic approach. ACS Appl. Nano Mater. 1: 2503–2508. https://doi.org/10.1021/acsanm.8b00559.

Mauter, M.S. and Elimelech, M. 2008. Environmental applications of carbon-based nanomaterials. Environ. Sci. Technol. 42: 5843–5859. https://doi.org/10.1021/es8006904.

Mercado, B.Q., Beavers, C.M., Olmstead, M.M., Chaur, M.N., Walker, K., Holloway, B.C. et al. 2008. Is the isolated pentagon rule merely a suggestion for endohedral fullerenes? The Structure of a Second Egg-Shaped Endohedral Fullerene—$Gd_3N@C_s(39663)$-C_{82}. J. Am. Chem. Soc. 130: 7854–7855. https://doi.org/10.1021/ja8032263.

Meyer, J.C., Geim, A.K., Katsnelson, M.I., Novoselov, K.S., Booth, T.J. and Roth, S. 2007. The structure of suspended graphene sheets. Nature 446: 60–63. https://doi.org/10.1038/nature05545.

Minami, K., Okamoto, K., Harano, K., Noiri, E. and Nakamura, E. 2018. Hierarchical assembly of siRNA with tetraamino fullerene in physiological conditions for efficient internalization into cells and knockdown. ACS Appl. Mater. Interfaces 10: 19347–19354. https://doi.org/10.1021/acsami.8b01869.

Mochalin, V.N., Shenderova, O., Ho, D. and Gogotsi, Y. 2012. The properties and applications of nanodiamonds. Nat. Nanotechnol. 7: 11–23. https://doi.org/10.1038/nnano.2011.209.

Mody, V., Siwale, R., Singh, A. and Mody, H. 2010. Introduction to metallic nanoparticles. J. Pharm. Bioallied Sci. 2: 282. https://doi.org/10.4103/0975-7406.72127.

Moreno-Vega, A.-I., Gómez-Quintero, T., Nuñez-Anita, R.-E., Acosta-Torres, L.-S. and Castaño, V. 2012. Polymeric and ceramic nanoparticles in biomedical applications. J. Nanotechnol. 2012: 1–10. https://doi.org/10.1155/2012/936041.

Morozov, S.V., Novoselov, K.S., Katsnelson, M.I., Schedin, F., Elias, D.C., Jaszczak, J.A. et al. 2008. Giant intrinsic carrier mobilities in graphene and its bilayer. Phys. Rev. Lett. 100: 016602. https://doi.org/10.1103/PhysRevLett.100.016602.

Nascimento, M.A., Cruz, J.C., Rodrigues, G.D., de Oliveira, A.F. and Lopes, R.P. 2018. Synthesis of polymetallic nanoparticles from spent lithium-ion batteries and application in the removal of reactive blue 4 dye. J. Clean. Prod. 202: 264–272. https://doi.org/10.1016/j.jclepro.2018.08.118.

Ng, K.K. and Zheng, G. 2015. Molecular interactions in organic nanoparticles for phototheranostic applications. Chem. Rev. 115: 11012–11042. https://doi.org/10.1021/acs.chemrev.5b00140.

Novoselov, K.S., Geim, A.K., Morozov, S.V., Jiang, D., Katsnelson, M.I., Grigorieva, I.V. et al. 2005. Two-dimensional gas of massless Dirac fermions in graphene. Nature 438: 197–200. https://doi.org/10.1038/nature04233.

Novoselov, K.S., Geim, A.K., Morozov, S.V., Jiang, D., Zhang, Y., Dubonos, S.V. et al. 2004. Electric field effect in atomically thin carbon films. Science. 306: 666–669. https://doi.org/10.1126/science.1102896.

Novoselov, K.S., Jiang, D., Schedin, F., Booth, T.J., Khotkevich, V.V., Morozov, S.V. et al. 2005 Two-dimensional atomic crystals. Proc. Natl. Acad. Sci. 102: 10451–10453. https://doi.org/10.1073/pnas.0502848102.

Novoselov, K.S., Morozov, S.V., Mohinddin, T.M.G., Ponomarenko, L.A., Elias, D.C., Yang, R. et al. 2007. Electronic properties of graphene. Phys. Status Solidi 244: 4106–4111. https://doi.org/10.1002/pssb.200776208.

Oh, W.-K., Yoon, H. and Jang, J. 2010. Size control of magnetic carbon nanoparticles for drug delivery. Biomaterials 31: 1342–1348. https://doi.org/10.1016/j.biomaterials.2009.10.018.

Ohashi, Y., Koizumi, T., Yoshikawa, T., Hironaka, T. and Shiiki, K. 1997. Size effect in the in-plane electrical resistivity of very thin graphite crystals. TANSO 1997: 235–238. https://doi.org/10.7209/tanso.1997.235.

Okada, H., Komuro, T., Sakai, T., Matsuo, Y., Ono, Y., Omote, K. et al. 2012. Preparation of endohedral fullerene containing lithium (Li@C60) and isolation as pure hexafluorophosphate salt ([Li+@C60][PF6−]). RSC Adv. 2: 10624. https://doi.org/10.1039/c2ra21244g.

Pachidis, P., Cote, B.M. and Ferry, V.E. 2019. Tuning the polarization and directionality of photoluminescence of achiral quantum dot films with chiral nanorod dimer arrays: implications for luminescent applications. ACS Appl. Nano Mater. 2: 5681–5687. https://doi.org/10.1021/acsanm.9b01198.

Pan, K. and Zhong, Q. 2016. Organic nanoparticles in foods: fabrication, characterization, and utilization. Annu. Rev. Food Sci. Technol. 7: 245–266. https://doi.org/10.1146/annurev-food-041715-033215.

Pan, Y., Guo, Z., Ran, S. and Fang, Z. 2020. Influence of fullerenes on the thermal and flame-retardant properties of polymeric materials. J. Appl. Polym. Sci. 137: 47538. https://doi.org/10.1002/app.47538.

Patolsky, F., Timko, B.P., Yu, G., Fang, Y., Greytak, A.B., Zheng, G. et al. 2006. Detection, stimulation, and inhibition of neuronal signals with high-density nanowire transistor arrays. Science. 313: 1100–1104. https://doi.org/10.1126/science.1128640.

Pokutnyi, S.I. 2003. Interband absorption of light in semiconductor nanostructures. Semiconductors 37: 718–722. https://doi.org/10.1134/1.1582542.

Popov, A.A., Yang, S. and Dunsch, L. 2013. Endohedral fullerenes. Chem. Rev. 113: 5989–6113. https://doi.org/10.1021/cr300297r.

Popov, V. 2004. Carbon nanotubes: properties and application. Mater. Sci. Eng. R Reports 43: 61–102. https://doi.org/10.1016/j.mser.2003.10.001.

Rafiee, M.A., Yavari, F., Rafiee, J. and Koratkar, N. 2011. Fullerene–epoxy nanocomposites-enhanced mechanical properties at low nanofiller loading. J. Nanoparticle Res. 13: 733–737. https://doi.org/10.1007/s11051-010-0073-5.

Rao, C.N.R., Satishkumar, B.C., Govindaraj, A. and Nath, M. 2001. Nanotubes. ChemPhysChem 2: 78–105. https://doi.org/10.1002/1439-7641(20010216)2:2<78::AID-CPHC78>3.0.CO;2-7.

Rauscher, H., Sokull-Klüttgen, B. and Stamm, H. 2013. The European Commission's recommendation on the definition of nanomaterial makes an impact. Nanotoxicology 7: 1195–1197. https://doi.org/10.3109/17435390.2012.724724.

Rodríguez-Fortea, A., Balch, A.L. and Poblet, J.M. 2011. Endohedral metallofullerenes: a unique host–guest association. Chem. Soc. Rev. 40: 3551. https://doi.org/10.1039/c0cs00225a.

Rother, M., Brohmann, M., Yang, S., Grimm, S.B., Schießl, S.P., Graf, A. et al. 2017. Aerosol-jet printing of polymer-sorted (6,5) carbon nanotubes for field-effect transistors with high reproducibility. Adv. Electron. Mater. 3: 1700080. https://doi.org/10.1002/aelm.201700080.

Sahoo, N.G., Rana, S., Cho, J.W., Li, L. and Chan, S.H. 2010. Polymer nanocomposites based on functionalized carbon nanotubes. Prog. Polym. Sci. 35: 837–867. https://doi.org/10.1016/j.progpolymsci.2010.03.002.

Sánchez-López, E., Gomes, D., Esteruelas, G., Bonilla, L., Lopez-Machado, A.L., Galindo, R. et al. 2020. Metal-based nanoparticles as antimicrobial agents: an overview. Nanomaterials 10: 292. https://doi.org/10.3390/nano10020292.

Sánchez, L., Martín, N. and Guldi, D.M. 2005. Hydrogen-bonding motifs in fullerene chemistry. Angew. Chemie Int. Ed. 44: 5374–5382. https://doi.org/10.1002/anie.200500321.

Sandhu, A. 2006. Who invented nano? Nat. Nanotechnol. 1: 87–87. https://doi.org/10.1038/nnano.2006.115.

Sathyanarayanan, M.B., Balachandranath, R., Genji Srinivasulu, Y., Kannaiyan, S.K. and Subbiahdoss, G. 2013. The effect of gold and iron-oxide nanoparticles on biofilm-forming pathogens. ISRN Microbiol. 2013: 1–5. https://doi.org/10.1155/2013/272086.

Schroer, Z.S., Wu, Y., Xing, Y., Wu, X., Liu, X., Wang, X. et al. 2019. Nitrogen–sulfur-doped graphene quantum dots with metal ion-resistance for bioimaging. ACS Appl. Nano Mater. 2: 6858–6865. https://doi.org/10.1021/acsanm.9b01309.

SCOTT, N.R. 2005. Nanotechnology and animal health. Rev. Sci. Tech. l'OIE 24: 425–432. https://doi.org/10.20506/rst.24.1.1579.

Shen, C., Brozena, A.H. and Wang, Y. 2011. Double-walled carbon nanotubes: Challenges and opportunities. Nanoscale 3: 503–518. https://doi.org/10.1039/C0NR00620C.

Shiman, D.I., Sayevich, V., Meerbach, C., Nikishau, P.A., Vasilenko, I.V., Gaponik, N. et al. 2019. Robust polymer matrix based on isobutylene (co)polymers for efficient encapsulation of colloidal semiconductor nanocrystals. ACS Appl. Nano Mater. 2: 956–963. https://doi.org/10.1021/acsanm.8b02262.

Shirasaki, Y., Supran, G.J., Bawendi, M.G. and Bulović, V. 2013. Emergence of colloidal quantum-dot light-emitting technologies. Nat. Photonics 7: 13–23. https://doi.org/10.1038/nphoton.2012.328.

Shnoudeh, A.J., Hamad, I., Abdo, R.W., Qadumii, L., Jaber, A.Y., Surchi, H.S. et al. 2019. Synthesis, characterization, and applications of metal nanoparticles. pp. 527–612. *In*: Biomaterials and Bionanotechnology. Elsevier. https://doi.org/10.1016/B978-0-12-814427-5.00015-9.

Shokuhi Rad, A. and Ayub, K. 2017. Substitutional doping of zirconium-, molybdenum-, ruthenium-, and palladium: An effective method to improve nonlinear optical and electronic property of C20 fullerene. Comput. Theor. Chem. 1121: 68–75. https://doi.org/10.1016/j.comptc.2017.10.015.

Singh, V.K.M., Yadav, S., Mishra, H., Kumar, R., Tiwari, R.S., Pandey, A. et al. 2019. WS 2 quantum dot graphene nanocomposite film for UV photodetection. ACS Appl. Nano Mater. 2: 3934–3942. https://doi.org/10.1021/acsanm.9b00820.

Song, J., Yu, Z., Gordin, M.L. and Wang, D. 2016. Advanced sulfur cathode enabled by highly crumpled nitrogen-doped graphene sheets for high-energy-density lithium–sulfur batteries. Nano Lett. 16: 864–870. https://doi.org/10.1021/acs.nanolett.5b03217.

Specification, T. and Preview, T.S. 2000. Technical Specification ISO/TS iTeh Standard Preview iTeh Standard Preview 2000.

Suh, Y.-H., Kim, T., Choi, J.W., Lee, C.-L. and Park, J. 2018. High-performance $CsPbX_3$ perovskite quantum-dot light-emitting devices via solid-state ligand exchange. ACS Appl. Nano Mater. 1: 488–496. https://doi.org/10.1021/acsanm.7b00212.

Takabayashi, Y. and Prassides, K. 2016. Unconventional high- T c superconductivity in fullerides. Philos. Trans. R. Soc. A Math. Phys. Eng. Sci. 374: 20150320. https://doi.org/10.1098/rsta.2015.0320.

Tang, J., Huo, Z., Brittman, S., Gao, H. and Yang, P. 2011. Solution-processed core–shell nanowires for efficient photovoltaic cells. Nat. Nanotechnol. 6: 568–572. https://doi.org/10.1038/nnano.2011.139.

Tiwari, S.K., Kumar, V., Huczko, A., Oraon, R., Adhikari, A. De. and Nayak, G.C. 2016. Magical allotropes of carbon: prospects and applications. Crit. Rev. Solid State Mater. Sci. 41: 257–317. https://doi.org/10.1080/10408436.2015.1127206.

Tiwari, S.K., Mishra, R.K., Ha, S.K. and Huczko, A. 2018. Evolution of graphene oxide and graphene: from imagination to industrialization. ChemNanoMat 4: 598–620. https://doi.org/10.1002/cnma.201800089.

Toshima, N. and Yonezawa, T. 1998. Bimetallic nanoparticles—novel materials for chemical and physical applications. New J. Chem. 22: 1179–1201. https://doi.org/10.1039/a805753b.

Tuktarov, A.R., Khuzin, A.A. and Dzhemilev, U.M. 2020. Fullerene-containing lubricants: achievements and prospects. Pet. Chem. 60: 113–133. https://doi.org/10.1134/S0965544120010144.

Valizadeh, A., Mikaeili, H., Samiei, M., Farkhani, S.M., Zarghami, N., Kouhi, M. et al. 2012. Quantum dots: synthesis, bioapplications, and toxicity. Nanoscale Res. Lett. 7: 480. https://doi.org/10.1186/1556-276X-7-480.

Wagner, R.S. and Ellis, W.C. 1964. Vapor-liquid-solid mechanism of single crystal growth. Appl. Phys. Lett. 4: 89–90. https://doi.org/10.1063/1.1753975.

Wei, Z., Kondratenko, M., Dao, L.H. and Perepichka, D.F. 2006. Rectifying diodes from asymmetrically functionalized single-wall carbon nanotubes. J. Am. Chem. Soc. 128: 3134–3135. https://doi.org/10.1021/ja053950z.

Wu, Y., Fan, R. and Yang, P. 2002. Block-by-block growth of single-crystalline Si/SiGe superlattice nanowires. Nano Lett. 2: 83–86. https://doi.org/10.1021/nl0156888.

Wu, Y. and Yang, P. 2001. Direct observation of vapor−liquid−solid nanowire growth. J. Am. Chem. Soc. 123: 3165–3166. https://doi.org/10.1021/ja0059084.

Xiao, J.R., Gama, B.A. and Gillespie, J.W. 2005. An analytical molecular structural mechanics model for the mechanical properties of carbon nanotubes. Int. J. Solids Struct. 42: 3075–3092. https://doi.org/10.1016/j.ijsolstr.2004.10.031.

Yan, R., Gargas, D. and Yang, P. 2009. Nanowire photonics. Nat. Photonics 3: 569–576. https://doi.org/10.1038/nphoton.2009.184.

Yang, Y., Liu, R., Wu, J., Jiang, X., Cao, P., Hu, X. et al. 2015. Bottom-up fabrication of graphene on silicon/silica substrate via a facile soft-hard template approach. Sci. Rep. 5: 13480. https://doi.org/10.1038/srep13480.

Yoshida, Z. and Osawa, E. 1971. Aromaticity, Chemical Monograph Series 22, Chemical Monograph Series 22. Kagaku-dojin, Kyoto.

Yousfi-Steiner, N., Moçotéguy, P., Candusso, D. and Hissel, D. 2009. A review on polymer electrolyte membrane fuel cell catalyst degradation and starvation issues: Causes, consequences and diagnostic for mitigation. J. Power Sources 194: 130–145. https://doi.org/10.1016/j.jpowsour.2009.03.060.

Yuan, W.Z., Mao, Y., Zhao, H., Sun, J.Z., Xu, H.P., Jin, J.K. et al. 2008. Electronic interactions and polymer effect in the functionalization and solvation of carbon nanotubes by pyrene- and ferrocene-containing poly(1-alkyne)s. Macromolecules 41: 701–707. https://doi.org/10.1021/ma701956a.

Yuan, X., Zhang, X., Sun, L., Wei, Y. and Wei, X. 2019. Cellular toxicity and immunological effects of carbon-based nanomaterials. Part. Fibre Toxicol. 16: 18. https://doi.org/10.1186/s12989-019-0299-z.

Zhan, G.-D., Kuntz, J.D., Wan, J. and Mukherjee, A.K. 2003. Single-wall carbon nanotubes as attractive toughening agents in alumina-based nanocomposites. Nat. Mater. 2: 38–42. https://doi.org/10.1038/nmat793.

Zhang, W.-D. and Zhang, W.-H. 2009. Carbon nanotubes as active components for gas sensors. J. Sensors 2009: 1–16. https://doi.org/10.1155/2009/160698.

Zhang, X., Ma, Y., Fu, S. and Zhang, A. 2019. Facile synthesis of water-soluble fullerene (C60) nanoparticles via mussel-inspired chemistry as efficient antioxidants. Nanomaterials 9: 1647. https://doi.org/10.3390/nano9121647.

Zhang, Y., Tan, Y.-W., Stormer, H.L. and Kim, P. 2005. Experimental observation of the quantum Hall effect and Berry's phase in graphene. Nature 438: 201–204. https://doi.org/10.1038/nature04235.

Zhao, Y., Li, X., Zhou, X. and Zhang, Y. 2016. Review on the graphene based optical fiber chemical and biological sensors. Sensors Actuators B Chem. 231: 324–340. https://doi.org/10.1016/j.snb.2016.03.026.

CHAPTER 2

Potential of Nanocellulose-based Aerogels for Industrial Use

Mahmood Ul Hassan,[1,*]
Saddam Hussain[2] and *Quanzhen Wang*[1]

1. Introduction

Nanocellulose is a desirable and sustainable material for aerogel (Pottathara et al. 2018) applications as it is derived from readily available resources, is inexpensive and renewable, and can be commercially produced. Current research is focused on discovering and implementing new uses for cellulose aerogels. The distinctive qualities of these items are: its wide availability, low density, high form factor, biological source of origin and nano-size. The substance nanocellulose is derived from wood fibers. Nanocellulose (also known as micro-fibrillated cellulose, MFC, or nano-fibrillated cellulose, NFC) has existed since the early 1980s. The primary components of plant fibers are cellulose, hemicellulose and lignin. Cellulose is the primary constituent of numerous natural fibers (Iwamoto et al. 2007), including cotton, flax, hemp, jute and sisal, among others. This natural polymer comprises approximately one-third of plant tissues and can be replenished through photosynthesis. Nanocellulose has a significantly smaller diameter (usually 10 nm) compared to cellulose, the fundamental component of the paper we use every day.

Thus, nanoscale cellulose exhibits unique physical and chemical features that make it an attractive material for ecologically friendly

---

[1] College of Grassland Agriculture, Northwest A&F University, Yangling, Shaanxi, 712100, China.
[2] Department of Agronomy, University of Agriculture, Faisalabad, 38000, Pakistan.
Email: wangquanzhen191@163.com
[*] Corresponding author: mahmood@nwafu.edu.cn

products. Extraction of cellulose nanofibers (CNFs) from waste biomass by means of steam explosion (Cherian et al. 2010) is accompanied by moderate acid hydrolysis (Pelissari et al. 2014) and high-pressure homogenization (Pääkko et al. 2007). Aerogels are produced through a technique known as freeze drying (Korhonen et al. 2011). Nanocellulose aerogels have a high absorption ability because of their large surface area and high number of active sites for interaction, as observed in numerous publications. Scientists are currently concentrating on the application of nanocellulose aerogels in oil/water separation, thermal insulation (Jiménez-Saelices et al. 2017), air purifiers (Nemoto et al. 2015), drug delivery (Fleischmann et al. 2020), shape memory applications (Gu et al. 2018), pressure sensing (Muthuraj et al. 2015), wound dressing, etc.

2. Preparation of Nanocellulose Aerogels

The microstructures and properties of nanocellulose aerogels can be influenced by cellulose supply and the manufacturing technique. Typically, aerogels derived from nanocellulose are produced in three steps:

(i) Nanocellulose dispersion
(ii) Nanocellulose gelaton
(iii) Gel drying method

2.1 Dispersion and Gelation of Nanocellulose

Nanocellulose is susceptible to intermolecular and intramolecular hydrogen bonds (Klemm et al. 2011) coupled with self-aggregation and tangling of nanocrystals or nanofibrils, due to the presence of a significant number of active hydroxyl groups on its surface. Due to the electrostatic repulsion between negatively charged nanocellulose (Sun et al. 2017), such as 2,2,6,6-tetramethylpiperidine-1-oxyradical (TEMPO)-oxidized CNFs and sulfonated CNCs, certain negatively charged groups (Nechyporchuk et al. 2016) can be incorporated onto the nanocellulose surface to create a stable and consistent aqueous nanocellulose dispersion. When CNFs are disseminated in water, hydrogen bonding and entanglement of the long fibrils can form a three-dimensional network.

As a network backbone, these nanofibrils can enhance the gel's strength and modulus of cellulose. Consequently, the gel does not contract substantially during the subsequent drying process. This allows the manufactured aerogel to retain structural integrity despite a low nanocellulose solids content. However, such a modified nanocellulose cannot be effectively dispersed in low-polarity organic solvents (e.g., tert-butyl alcohol (TBA) or ethanol) which are widely employed in the aerogel drying process (Nemoto et al. 2015). Interestingly, the addition of a tiny quantity of TBA to the aqueous CNF dispersion can equally scatter nanofibrils in the mixed solution and increase the specific surface area (> 300 m^2g^{-1}) of the nanocellulose aerogel generated by freeze drying.

After acquiring the precursor dispersion, several nanocellulose aerogels must typically undergo a gelation process, which aids in the preservation and improvement of the 3D network within the aerogel. Depending on the nature of the gel, the gelation behavior of nanocellulose is commonly divided into two broad categories: chemical crosslinking and physical crosslinking (Zhang et al. 2019). Chemical crosslinking entails the addition of certain crosslinking agents (multifunctional monomers) to a solution, like citric acid or glutaraldehyde (Medina et al. 2019) which react with nanocellulose to generate irreversible covalent links between the cellulose chains. Qu et al. (2018) used 1,2,3,4-butanetetracarboxylic acid (BTCA) as a chemical crosslinker to help TEMPO-oxidized CNFs in forming a crosslinking network structure in order to produce a mechanically resilient aerogel at an extraordinarily low concentration (0.3 wt%) of nanocellulose.

2.2 Gel Drying

After dispersal and gelation of nanocellulose to produce a stable three-dimensional network structure, it is important to replace the liquid solvent packed in the structure with gas, while preserving the original structure, to obtain an aerogel. The most crucial stage in preparing nanocellulose aerogel is selecting an appropriate drying method. Hence, various drying methods have been devised to create a porous aerogel network during solvent evaporation.

In general, the majority of the solvent in the network gel structure is water, which has a rather high surface tension. Due to the enormous capillary pressure (F1, F2, F3 and F4) and the resulting forces (F1′, F2′, F3′, F4′) exerted on pore walls in the horizontal direction between gas–liquid interfaces during solvent evaporation, the pore structure collapses and is destroyed. The capillary pressure (P) can be calculated using the Young–Laplace equation (Li et al. 2019):

$$P = \frac{-2\gamma \cos \theta}{r}$$

where P, r and θ represent the surface tension, the pore radius and the contact angle between the solvent and pore walls respectively. Consequently, there are a number of ways to reduce the capillary pressure:

i) Reducing the γ of the solvent.
ii) Increasing the pore size r to decrease the θ.

iii) Regulation of the uniformity of pore size to optimize the pressure on pore walls; and

iv) Improving the mechanical properties of the pore walls.

For direct and user-friendly atmospheric pressure drying, it is typically possible to replace water or alcohol in the nanocellulose gel with a solvent exhibiting low surface tension or to incorporate some oil-dissolving surfactants to reduce the surface tension, thereby reducing the internal capillary pressure. Likewise, Toivonen et al. (2015). demonstrated an atmospheric-pressure-dried CNF aerogel membrane produced by vacuum filtering and solvent swapping with 2-propanol and, subsequently, octane slow drying of the CNF gel in octane produced a CNF aerogel membrane with a thickness of 25 m, in contrast to the 12 m translucent and compacted CNF film formed by rewetting the CNF aerogel membrane with water and then drying it.

2.3 *Supercritical Drying*

Supercritical drying entails replacing the solvent in the nanocellulose gel with supercritical fluids due to their superior solubility, diffusivity and surface tension. Similar to a drying kettle, in this case as well, the ideal temperature and pressure are maintained to allow the solvent to achieve its critical point, followed by a gradual pressure release and cooling. By discharging the solvent from the liquid phase to the gaseous phase via supercritical transformation, a structurally complete aerogel is produced (Plappert et al. 2018). As the solvent exhibits no discernible surface tension during this transformation, the wet gel can be transformed into an aerogel while retaining its skeleton structure. Under supercritical drying, carbon dioxide (CO_2) and ethanol are frequently acceptable supercritical fluids for nanocellulose gels.

However, the critical temperature of ethanol is above 200°C, posing concerns associated with its use as a supercritical fluid, as well as difficulties in mass production (García-González et al. 2012). CO_2's critical temperature is only 31.3°C, which is slightly above room temperature, making it more stable. Therefore, it is considerably safer and more prevalent in use. A nanocellulose aerogel was produced by Zu et al. (2016) by swapping the precursor gel with ethanol, and supercritical CO_2 drying. When the gel liquid and CO_2 reach the supercritical state (31.3°C and 72.9 atm), gas–liquid conversion, a process of two-way mass transfer between the solvent in gels and the fluid, takes place. This substitution does not produce liquid surface tension, hence preventing gel structure collapse during the drying process. In addition, the acquired nanocellulose aerogel showcases an interconnected 3D network nanostructure with relatively more homogeneous pores and less shrinkage, and a larger SSA of 299 m^2g^{-1} than the freeze-dried sample (103 m^2g^{-1}), hence confirming that supercritical CO_2 drying is a more effective method than freeze drying for preserving the nanocellulose gel structure. However, the application of supercritical drying is restricted to high-quality products due to its complexity, longer manufacturing cycle and higher solvent replacement cost.

2.4 Freeze Drying

The most popular technique for removing the solvent from the nanocellulose gel and controlling the network topology within the aerogel to prevent collapse is freeze drying. Moreover, compared to the previous two processes, freeze drying offers environmental protection, high efficiency and low cost, thus garnering significant interest in nanocellulose aerogels.

Freeze drying is essentially a sublimation drying process:

i) It involves the freezing of the nanocellulose wet gel to a solid state.

ii) At a specific low temperature and pressure, the frozen solvent (mainly water) in the internal gel is directly sublimated from the solid state to the gaseous state.

iii) As the solvent is discharged, multiple pores are maintained to form a porous aerogel.

This procedure avoids gas–liquid phase interaction. Thus, it can effectively prevent the formation of capillary pressure during the drying process, hence preserving the structural integrity of the skeleton (Cheng et al. 2017). In general, the majority of nanocellulose aerogels are created by submerging the precursor dispersion/gel immediately in liquid nitrogen or refrigeration. After sublimation, the ultracold liquid nitrogen (196°C) causes the water solvent to rapidly form ice crystals, leaving dense and minute holes in the aerogel. Notably, employing a cold source with a greater temperature requires longer times to develop larger holes due to the slower freezing rate (Mueller et al. 2015).

The nanocellulose aerogels mentioned above have pores of varying diameters, 20 m at 196°C for 5 minutes and 200 m at 26°C for 24 hours. Nevertheless, the microporous structure of conventional aerogels is often isotropic, which is a disadvantage of the conventional freezing process. The disorganized structure of aerogel prohibits it from performing activities such as directional mass transport, heat transmission and electrical conductivity. Due to its lengthy and winding transmission channel, this material has limited applicability. In addition, the disordered nanocellulose aerogel is unmanageable in comparison to the 3D organized structure. For functional structural materials to carry out their functions, proper regulation and assembly of the micro/nanoscale structure is crucial.

On the basis of the conventional freeze drying method for successfully managing the pore structure of ice crystal-containing materials, directional freeze drying (primarily consisting of unidirectional freeze drying and bidirectional freeze drying) was developed. Compared to the conventional processes for creating porous materials, it is recognized to have various advantages, including simplicity, adaptability and a wide range of applications (e.g., particle leaching, foaming and phase separation) (Chen et al. 2019). During anisotropic freezing, the temperature gradient is delivered in only one or two directions of the material, as opposed to all directions during isotropic freezing. After applying a bottom-up temperature gradient for unidirectional freeze drying, the ice crystals of the solvent grow in this direction. During this procedure, the solute (CNFs) in the solution is extruded on to the interfaces between ice crystals to achieve a solid–liquid separation, followed by sublimation and drying with a drier to generate a CNF aerogel after full freezing.

The structural anisotropy is expressed in the hexagonal honeycomb holes in the transverse direction and aligned ordered directional tunnels in the longitudinal direction. Moreover, tilting the bottom of the container containing the precursor at a certain angle (20°) allows the formation of two temperature gradients along the Y-axis and Z-axis due to contact with the cold source liquid nitrogen, resulting in the formation of a bidirectional anisotropic aerogel after drying.

Due to their well-aligned lamellar structures, these unique aerogels demonstrate thermal insulation and mechanical capabilities superior to isotropic and unidirectional aerogels (Zhang et al. 2020). Employing an appropriate cold source (liquid nitrogen at 196°C or milder cold ethanol with adjustable temperatures) and temperature gradients, the selective adjustment of the aerogel structure can be achieved using this simple and flexible directional freeze drying technique; this has been reinforced in various fields of nanocellulose aerogel applications, including selective absorption (Yang et al. 2018), gas capture (Geng et al. 2020), solar steam generation, EMI shielding (Zeng et al. 2020) and energy storage.

Customizable nanocellulose aerogels with finely regulated 3D architectures and inner pore architecture have been recently created using a 3D printing method (Lei et al. 2021). Using the 3D printing approach can compensate for the inability of the aforementioned-methods to rapidly produce aerogels with complicated shapes, hence broadening the applicability of nanocellulose aerogels to fulfill the unique requirements of numerous applications, including tissue engineering (Li et al. 2017).

3. Extraction of Cellulose Nanofibers (CNFs) and Cellulose Nanocrystals (CNCs)

Nanocelluloses are primarily categorized as (1) cellulose nanofibers (CNFs) and (2) cellulose nanocrystals (CNCs). CNCs have nano dimensions in both length and diameter, whereas the CNFs have micro dimensions in length and nano dimensions in diameter. Isolation of nanocellulose from renewable sources is gaining importance amongst scientists worldwide. Numerous techniques for extracting CNFs from the plant cell wall have been documented; several mechanical processes have been explored to extract CNFs from a variety of sources as well. In addition to high-pressure homogenization, grinding, cryocrushing and high-intensity ultrasonic treatment, which results in the transverse splitting of the cellulose fibers, sulfuric acid has been routinely employed to extract CNCs from cellulose materials. Recently, Danial et al. recovered CNCs from waste paper through acid hydrolysis, using a 60% (V/V) sulfuric acid solution at 45°C with continuous stirring. They developed CNCs ranging from 100 to 300 nm in length (Danial 2015).

4. Comparison between Nanocellulose Aerogels and Papers

It has been demonstrated that nanocellulose may be treated into many forms, such as aerogels, sheets and hydrogels, among others. Due to their ease of production and broad range of applications, nanocellulose aerogel and paper have attracted a great deal of interest from the scientific world as a whole. Aerogels are ultralow-density porous structures with

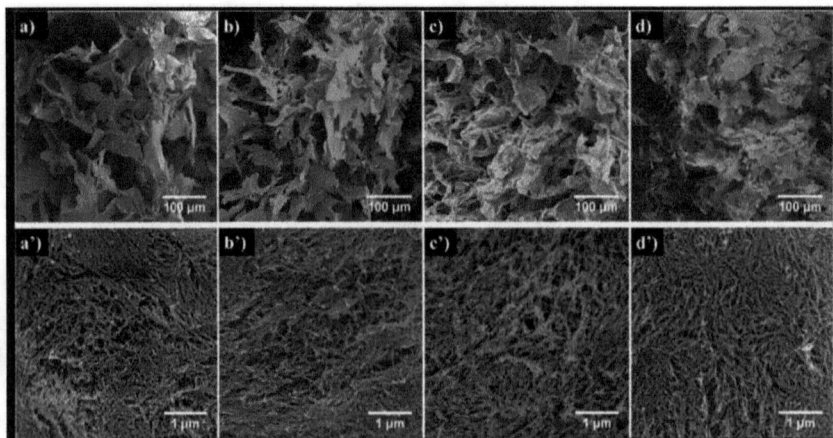

Figure 1: SEM images of aerogels with different starting CNC concentrations: (a, a') 0.5 wt %, (b, b') 1.0 wt %, (c, c') 1.5 wt %, and (d, d') 2.0 wt %. Macropores are visible at low magnifications (top, a–d) and mesopore uniformity is observed as higher.

exceptional mechanical qualities; this makes them fit for a variety of applications, such as biomedical scaffolds, thermal insulation, and energy storage and generating devices. Traditionally, nanocellulose aerogels have been manufactured using the freeze-drying technique, which has been employed by a large number of scientists throughout the world.

5. Applications of Nanocellulose

5.1 Nanocellulose as Super Disintegrants

Nanocellulose, often known as 'cellulose whiskers', may play a crucial role in the pharmaceutical industry. They function as disintegrants in the production of fast-dissolving, orally disintegrating and mouth-dissolving tablets (Sheikhy et al. 2021). The inherent structure of cellulose encompasses both amorphous and crystalline cellulose units. After acid hydrolysis, cellulose often takes on a highly crystalline structure, while the amorphous portion is also retained. Nanocellulose has replaced microcrystalline materials due to its poor swelling capacity and good porosity (Yassin et al. 2015). Numerous sources demonstrate that nanocellulose can be subdivided into freeze-dried and spray-dried nanofibrils and nanocrystals, all of which are effective super disintegrants. Cellulose nanofibrils are superior to nanocrystals as a direct compressing agent. Using nanocellulose as super disintegrants allows for a variety of solid dosage form formulations.

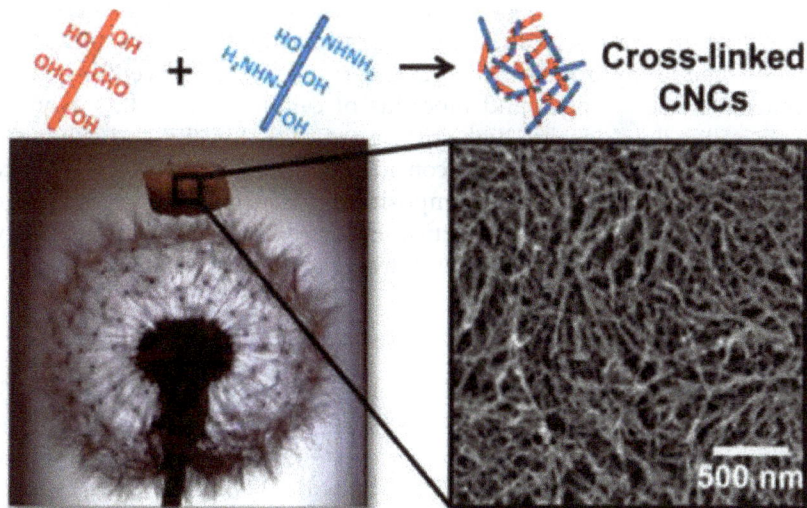

Figure 2: A low-light photograph of a piece of CNC aerogel (prepared from a 0.5 wt% CNC suspension) standing on the top of a dandelion flower. Copyright (2014). Reproduced with permission from American Chemical Society.

5.2 *Nanocellulose in Cosmetic Preparations*

Nanocellulose derived from bacterial cellulosic sources has numerous aesthetic applications. As an anti-wrinkle, thickening, stabilizing and coating agent, Nano Bacterial Cellulosic Fibers (NBCF) have multiple applications. The anti-wrinkle effect of these substances is a result of their tissue-healing properties; as stabilizing and thickening agents, they have a 'non-dipping effect'. Therefore, they can be employed in the formulation of sunscreen sprays, and also have coating properties that aid in the manufacture of products for nails, hair, eyebrows and eyelashes. Nanocellulose derived from bacteria is the most commonly used ingredient in cosmetics. Its 3D nanofibrous structure provides exceptional mechanical strength, flexibility and skin or topical route adhesion. Multiple studies indicate that Bacterial Nanocellulose (BNC) can be utilized in plastic surgery to heal scars and eliminate edema.

BNC has a minor brightening effect (Ludwicka et al. 2016) as well. Cellulosic nanocrystals can be used to produce liquid foundations that cleanse and remove makeup without damaging the skin (Tang et al. 2021). Bacterial nanofibers can be used to create washes and face mask sheets (Almeida et al. 2021).

5.3 *Nanocellulose Hydrogels in Pharmaceutical Preparations*

Hydrogels manufactured from nanocellulose are highly porous and pliable materials with excellent mechanical properties (Curvello et al. 2019). By fusing the functional groups of cellulose in nanocrystals, fibrils and bacterio-nano cellulose, various properties and functionalities can be conferred on hydrogels. Hydrogels derived from nano cellulose have excellent tensile strength and modulus of cellulose as well as a broad aseptic range, low density, high surface area, low toxicity and enhanced optical characteristics. They are considered as 'green composites' since they are biodegradable. Nanocomposites hydrogels have a wide range of applications in the pharmaceutical industry, including drug delivery systems, wound healing and wound dressing (Nicu et al. 2021).

5.4 *Nanocellulose in Transdermal Drug Delivery System*

The Transdermal Drug Delivery System (TDDS) is the most lauded of the existing drug delivery systems due to its cutting-edge properties such as minimal pain of application, prevention of first-pass metabolism, prolonged action and unique ability to provide therapeutic efficacy with the fewest dosages. In this method, nanocellulose plays a vital function as the skin tissue heals. Using bacterial cellulose sources and chitosan cellulose sheets can satisfy the notion of nanocellulose in a transdermal medication delivery strategy.

5.5 Applications of Nanocellulose Ophthalmology

Nanocellulose is used in the production of ophthalmic contact lenses. Using polyvinyl alcohol and cellulose nanofibrils, a macroporous and highly transparent hydrogel containing more than 90 percent water has been created by researchers. The refractive index of the said hydrogel was close to that of water, with UV-blocking characteristics and flexibility like that of collagen (Tummala et al. 2017). Nano-cellulose plays a significant role in the industrial sector in functionalizing processes such as ultrafiltration, storage and packaging.

5.6 Ultrafiltration

Due to its smaller size, nanocellulose plays a function in filtration and ultrafiltration processes. They are more commonly deployed in the ultrafiltration procedure. Using nano-cellulose in paper or membrane form, also known as UF-nano-cellulosic membranes or sheets, filtration can be accomplished. These nano-cellulosic-based papers or membranes are fabricated from bacterial cellulose, nanocellulose obtained from wood, TEMPO-oxidized cellulose nanofibrils and cellulose nanocrystals. The first technique modification of selected polymer matrices is the incorporation of this nanocellulose for better performance of that polymeric membrane – in the method of film casting the polymer matrix must be completely dissolved in a solvent and there must be complete dispersion of the nano-cellulosic solution. The second process is solvent casting, which results in the highest porosities (Hassan et al. 2020).

5.7 Storage

Nanocellulose can function as an electrochemical energy storage device (EED) (Chen and Hu 2018). As a storage system, nano-cellulosic applications include super capacitors, lithium-ion batteries and post-lithium-ion batteries. Producing nanocrystal cellulose, nanofibril cellulose and bacterial cellulose forms is possible from nanocellulose. Their 1D structure and compatibility has resulted in enhanced engineering performance as a superior storage system compared to other typical synthetic devices (Kim et al. 2019).

5.8 Packaging

Diverse sources indicate that in the last few decades, nanocellulose has played an increasingly important role in the field of packaging, which is the proper hand process of industries– nanocellulose is not only eco-friendly, but also edible, flexible, biodegradable, transparent, rheological, thermomechanical and antimicrobial (Ahankari et al. 2021). Nanocellulose is a material suited for all sorts of packaging, including active packaging, controlled release packaging and responsive packaging, as evidenced by the literature. Due to its superior barrier characteristics, nanocellulose is an ideal material for packaging. Studies indicate that in addition to the other benefits, nanocellulose provides superior gas barrier (including water vapor barrier) and UV barrier qualities. In terms of gas barrier properties, the oxygen barrier is required for packaging as food items are susceptible to the growth of aerobic microorganisms which spoil food. So as to extend the shelf life of food, low oxygen levels must be maintained.

Owing to two factors, nanocellulose can be used as a packaging material. The first factor is the formation of hydrogen bonds by nanocellulose, which results in a network (capable of linking within and cross-linking with other polymers) that prevents the entry of gas without the involvement of minute pores; the second factor is nano-cellulosic crystallinity's opposition to gas. In terms of water vapor barrier properties, nano-cellulosic fibrils are the best contender for optimally acting as a water vapor barrier as they inhibit oxygen entry and optimize penetration in the film. When it comes to UV barrier properties, numerous studies indicate that nano-cellulosic crystal composites are superior antioxidants and antibacterials (Ahankari et al. 2021).

5.9 Tissue Engineering

Nanocellulose can be manufactured in a three-dimensional structure by a variety of processes that replicate the extracellular matrix required for specialized tissues. This idea is easily applicable to bone tissue engineering, cartilage tissue engineering, cell screening and cancer proliferation, among other applications. Recent research indicates that nano-cellulosic 3D-structured hydrogels can imitate the nanostructured collagen in the bone marrow with relative ease. Using a nano-cellulosic fibril model for skin tissue culture for dermal and epidermal infections, nano-cellulosic bilayer skin has been created by researchers (Subhedar et al. 2021).

5.10 Implants

A few studies indicate that nanocellulose, due to its biodegradable qualities, may be one of the candidates for implants. Different nanocellulose-containing polymeric compositions can thus serve as implants. Various vascular grafts, artificial blood vessels and gadgets have been produced based on these ideas, and additional research is also ongoing.

5.11 Applications of Biosensorics

Owing to the potential modification of its structure into the desired dimensions, nanocellulose can be used in the biosensing process due to its numerous features. In addition to the manufacturing of composites with metal oxides such as those of Cu and Au, alternative matrices containing fluorescent elements are also being developed by researchers. It is possible to create nano-cellulosic fibril, crystals and bacterial cellulose; using chemical treatment, grafting, coating, impregnation, covalent binding, cryo-crushing, micro-fluidization, high-intensity ultrasonication and other wet chemistry techniques, these types can be produced rapidly.

5.12 Miscellaneous

Nanocellulose offers a vast array of applications in numerous fields, with bacterial nanocellulose playing a crucial role in almost all approaches. Recent research indicates that the nano-cellulosic composite concept can be readily implemented in cardiovascular therapy.

6. Nanocellulose Aerogel for Varying Technological Applications

The research interest in aerogels derived from nanocellulose is new but expanding quickly. Nanocellulose provides a mix of qualities suited for the production of ultralight, robust and flexible foams and aerogels for an ever-expanding array of applications. The combination of ultralow density, customizable porous design and exceptional mechanical qualities makes them attractive for a variety of applications, such as biomedical scaffolds, thermal insulation, oil absorption, etc.

6.1 Nanocellulose Aerogel for Thermal Insulation Applications

Daily increases in global temperature have been attributed to global warming. The addition of thermal insulation to buildings is one of the ways to achieve a comfortable environment. The greenhouse effect has warmed the planet and endangered human life. In 2005, it was discovered that CO_2 emissions from buildings accounted for more than 30% of greenhouse gas emissions in developed nations. In 2021 buildings accounted for 39% of energy use (Norouzi et al. 2021). Increasing the thermal insulation of buildings is among the most efficient ways to address this problem. Therefore, there have been considerable efforts in inventing novel insulation materials. Numerous investigations and studies on the production and usability of nanocellulose aerogels for thermal insulation have been described in this regard.

For the construction of nanocellulose and graphene oxide-based foams, controlled freezing of an aqueous suspension of CNF, GO and SEP in a mold set on the top of a cold finger immersed in liquid nitrogen has been used by researchers (Deepu A. et al. 2021). Nguyen et al. (2013) developed a cost-effective and scalable method for producing biodegradable cellulose aerogels from paper waste. Therein, aerogels made from recycled cellulose were examined for their superior thermal insulation and absorption capabilities. It was demonstrated that the cellulose aerogel exhibited a thermal conductivity of 0.029–0.032 $Wm^{-1}K^{-1}$, using the thermal conductivity analyzer system. Its measured thermal conductivity was comparable to the thermal conductivities of good insulating materials such as silica aerogel (0.026 $Wm^{-1}K^{-1}$), wool (0.03–0.04 $Wm^{-1}K^{-1}$) and aspen aerogels products (0.021 $Wm^{-1}K^{-1}$). It was found that the recycled cellulose aerogel's low heat conductivity and low cost make it a potential material for thermal insulation applications (Nguyen et al. 2013).

6.2 Nanocellulose Aerogel for Flame Retardant Applications

Cellulose aerogels are ultralight three-dimensional (3D) porous materials with the excellent properties of highly porous aerogels (such as high specific surface area, low density, low thermal conductivity and high insulation), along with the fantastic properties of sustainable biopolymers (such as biodegradability and biocompatibility). Cellulose aerogel is considered one of the most promising heat-insulating materials due to its abundant internal pores and good thermal insulation performance. However, nanocellulose aerogels are easy to ignite, which is the main limitation to its wide applicability in many commercial applications. In this context, researchers have been more focused on developing flame retardant nanocellulose aerogels while maintaining their unique properties for varying technological applications. The prepared composite cellulose aerogels have shown excellent flame retardancy; the peak of heat release rate (PHRR) of the composite aerogels decreased significantly from 280 Wg f the control sample to 22 Wg and the total heat release (THR) of the composite aerogel decreased remarkably from 13.2 to 1.6 kJg (Jančič et al. 2021).

Moreover, the incorporation of AH NPs composite aerogels has yielded remarkable mechanical properties– the compressive strength of the composite aerogels increased significantly from 0.08 MPa to 1.5 MPa (Cen et al. 2023). In addition, AH NPs composite cellulose aerogels have excellent sound absorption at high frequencies, with a maximum sound absorption coefficient of 1. Thus, Nguyen et al. concluded that the demonstrated work was a facile strategy to rationally construct a flame retardant, mechanically robust, high-efficiency sound adsorbing and super amphiphobic cellulose-based composite aerogels which have promising applications in the future as green engineering materials.

In the study discussed above, the addition of silica nanoparticles enhanced the mesoporous properties of aerogels (particularly the specific surface area and the mesopore volume), greatly slowed down the degradation of cellulose and inhibited the heat release during burning. The composite aerogels with high silica content (at least 33.6%) demonstrated good transparency with light transmittance as high as 78.4% at 800 nm, much higher than pure cellulose aerogels, and presented outstanding flame retarding performance, self-extinguishing after igniting. The group believes that transparent cellulose-silica composite aerogels with higher mechanical efficiency and enhanced flame retardancy may have vast application potential. In addition, they discovered that the microporous cellulose gel generated from the NaOH/urea solution might act as a template for the non-agglomerated development of MH NPs.

6.3 Nanocellulose Aerogel for Biomedical Applications

Tissue engineering has played a significant role in the substitution of dysfunctional organs or tissue lost as a result of accidents or diseases. The engineering of polymer materials to replace broken or missing components has maintained its popularity. In recent years, replacement of organs and regeneration of tissue have garnered significant scientific interest. Tissue engineering is a multidisciplinary field that tries to create biological alternatives to restore, replace or enhance tissue functions. Different polymer implants and biomaterials have been investigated globally for tissue regeneration purposes. Polymers with superior mechanical characteristics and biocompatibility will play a significant role in the global medical business. Cells can attach and proliferate inside a bioactive environment created by polymer scaffolds.

Extensive research has been conducted on scaffolds composed of synthetic and natural biomaterials, as well as 3D scaffolds such as nanofibers, hydrogels and sintered micro particles. Tissue engineering is one of the most important application areas for nanocellulose. Due to its intriguing mechanical characteristics and biocompatibility, bacterial cellulose is intensively researched in the field of tissue engineering. They have exceptional water retention, and their highly fibrous network promotes cell development and proliferation. Researchers extracted nanocellulose from wood powder and oxidized it with sodium periodate to create dialdehyde NCFs. Additionally, the dialdehyde NCFs served as a matrix to direct the development of collagen and through chemical cross-linking, collagen grew along the dialdehyde NCFs to generate composite aerogels.

Liu et al. (2021) used presorption (Method I) and *in situ* adsorption (Method II) to incorporate various forms of hemicellulose (galactoglucomannan (GGM), xyloglucan (XG), and xylan) into NFC hydrogels. They discovered that XG had the maximum adsorption capacity on the NFC, the greatest reinforcing effect, and promoted cell proliferation. The presorbed XG in the low-charged NFC network with a lower weight ratio (NFC/XG-90:10) supported the growth and proliferation of fibroblast cells (NIH 3T3) most effectively.

6.4 Nanocellulose Aerogel for Oil Absorption

Large tanker oil spills are a source of oil pollution. However, there are other sources that collectively release more oil into the sea than these massive oil spills. When oil accumulates on the surface of water, there may be a variety of negative outcomes. The most prevalent impact is the spread of oil across the water surface. Most of the oils are less dense than water; hence, oil spills spread across the surface of the water. Oil spreads very swiftly, with lighter oils such as gasoline spreading faster than heavier crude oils. Wind and high temperatures also accelerate the spread of oil across the water surface. Oil pollution can harm ecosystems, including plants and animals, while also contaminating drinking water. Furthermore, oil can adhere to the feathers of birds and marine animals; when this occurs, these animals can no longer protect themselves against the cold water and birds have difficulties in flying.

Oil contamination can affect aquatic plants as well as plants that grow nearby. An oil spill can obstruct the sunlight required for photosynthesis, resulting in the death of aquatic vegetation. Moreover, oil pollution can impact fishing and hunting, which is especially devastating for those who rely on these activities. Sources of water designed for human consumption can also become contaminated. Because of this, it is crucial to clear oil spills from water. To this goal, research has been conducted to develop materials that do not absorb water but can swiftly absorb a huge amount of oil, are durable, floatable, reusable, eco-friendly, and can be burned along with the oil they absorb.

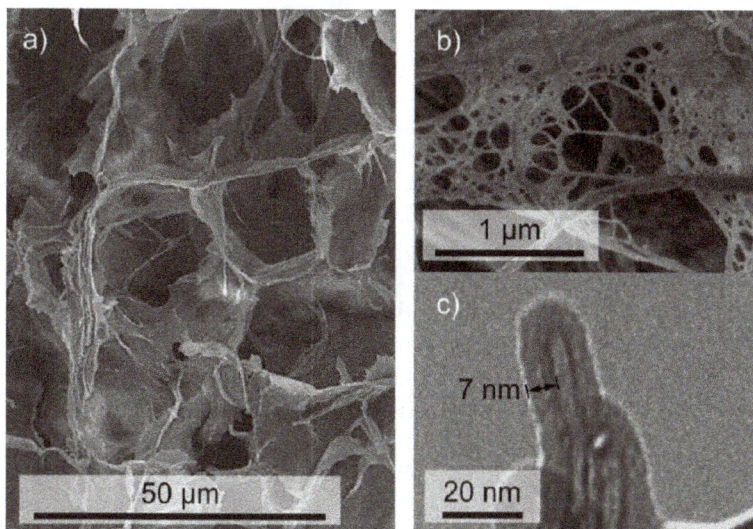

Figure 3: Microscopic structure of native nanocellulose aerogels. SEM micrographs of: (a) freeze-dried nanocellulose aerogels with fibrils packed into sheets, connected to form an open, porous aerogel structure, and (b) magnification of a sheet composed of fibrils making the structure nano-porous as well; (c) TEM micrograph of a nanocellulose fibril with a uniform 7 nm TiO_2 coating. Copyright (2011). Reproduced with permission from American Chemical Society.

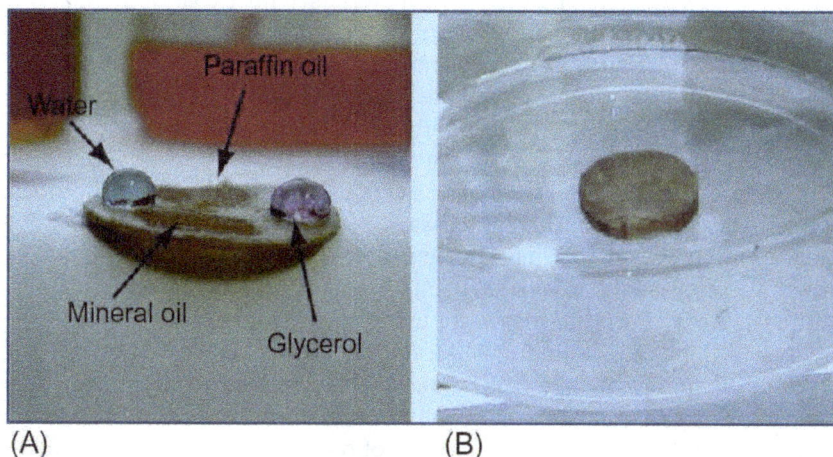

Figure 4: (A) TiO_2-coated aerogels are hydrophobic and oleophilic: water and glycerol stay as droplets (colored with reactive blue dye for clarity), whereas paraffin oil and mineral oil are readily absorbed. (B) Coated aerogel floating on water. Copyright (2011). Reproduced with permission from American Chemical Society.

Based on functionalized cellulose nanofibrils, Korhonen et al. (2011) demonstrated a material with cellulose as a natural, renewable and plentiful polymer. They created very porous nanocellulose aerogels by freeze-drying micro-fibrillated cellulose hydrogels under vacuum. By functionalizing the native cellulose nanofibrils of the aerogel with a hydrophobic titanium dioxide oleophilic covering, they demonstrated that the aerogel can preferentially absorb oils from water. Due to their low density and high capacity to absorb non-polar liquids and oils up to almost their entire initial volume, surface-modified aerogels can be used to collect organic pollutants from the water surface. It was demonstrated that the absorption was almost equal to the total volume of the aerogel (80%–90% vol/vol) and that the mass-based adsorption capacity fluctuated between 20 and 40 (wt/wt), depending on the liquid's density. It was also found that nano-cellulose-based aerogels would be an excellent contender for sustainable oil absorbent applications in the future.

7. Summary and Future Perspectives

In this chapter, cellulose-based aerogels and their industrial significance were discussed. The biocompatibility, biodegradability and low cost of aerogels make them a preferred choice for numerous industrial applications. Aerogel-based research in the future will concentrate mostly on the design of innovative materials with expanded functions and enhanced performance. This innovative design and architecture with cost-effective performance would increase the capacity of the concerned industrial applications. In the near future, cellulose-based composite aerogels with a specific design capability may replace the existing products for oil absorption, flame retardancy and biomedical applications. In the past decade, major efforts have been made to improve the performance of aerogels based on cellulose composites as a standalone component. The incorporation of nanoparticles or nanostructures in the form of aerogels into the cellulose matrix has been used to create multi-component products with improved characteristics and numerous functions.

In addition to facilitating the localization of nanostructures for target-level applications in biological domains, the porous nature and free space of aerogels protect them against deterioration. As a new class of innovative materials with the inclusion of nanostructures into an aerogel network with enhanced physicochemical, mechanical and thermal properties is highly suitable for a variety of industrial applications. Still, there is a need to clarify the connections between nanostructures and cellulose networks within aerogels, along with their mechanisms, in order to develop more scalable, cost-effective products for multipurpose applications. Future studies regarding the biological qualities of aerogels, such as *in vivo* circumstances and adsorption tests, require exhaustive examinations. In conclusion, a comprehensive research approach and practice should be built for studying the fundamental interactions and functions of aerogel systems designed in the future. In addition to these, effective collaboration between nanoscience, nanotechnology and medicine would be necessary to solve these challenges and develop materials with sturdy architectures. This chapter has selectively illuminated the industrial applications of cellulose-based aerogels in biomedical, flame-retarding, oil-absorption and thermal insulation applications.

8. Conclusion

Due to its singularity, a natural biopolymer is significant across multiple industries, particularly in medicine delivery systems. This review demonstrates the significance of natural biopolymers in many sectors. Through this chapter, the relevance of nanocellulose will be conveyed to the readers in a single location. Specifically, structured nanocellulose has a wider range of applications in oral drug delivery systems, whereas nanocellulose fibrils are mostly employed in transdermal drug delivery.

Acknowledgment

Mahmood Ul Hassan appreciates Dr. Saddam Hussain and Dr. Quanzhen Wang for their invaluable technical assistance, which significantly contributed to the development and quality of this chapter.

References

Ahankari, S.S., Subhedar, A.R., Bhadauria, S.S. and Dufresne, A. 2021. Nanocellulose in food packaging: A review. Carbohydr. Polym. https://doi.org/10.1016/j.carbpol.2020.117479.

Almeida, T., Silvestre, A.J.D., Vilela, C. and Freire, C.S.R. 2021. Bacterial nanocellulose toward green cosmetics: Recent progresses and challenges. Int. J. Mol. Sci. https://doi.org/10.3390/ijms22062836.

Cen, Q., Chen, S., Yang, D., Zheng, D. and Qiu, X. 2023. Full Bio-based aerogel incorporating lignin for excellent flame retardancy, mechanical resistance, and thermal insulation. ACS Sustain. Chem. Eng. 11: 4473–4484. https://doi.org/10.1021/acssuschemeng.2c07652.

Chen, C. and Hu, L. 2018. Nanocellulose toward advanced energy storage devices: Structure and electrochemistry. Acc. Chem. Res. 51: 3154–3165. https://doi.org/10.1021/acs.accounts.8b00391.

Chen, Y., Zhou, L., Chen, L., Duan, G., Mei, C. and Huang, C. 2019. Anisotropic nanocellulose aerogels with ordered structures fabricated by directional freeze-drying for fast liquid transport. Cellulose 26: 6653–6667. https://doi.org/10.1007/s10570-019-02557-z.

Cheng, Q., Huang, C. and Tomsia, A.P. 2017. Freeze casting for assembling bioinspired structural materials. Advanced Materials 29. https://doi.org/10.1002/adma.201703155.

Cherian, B.M., Leão, A.L., de Souza, S.F., Thomas, S., Pothan, L.A. and Kottaisamy, M. 2010. Isolation of nanocellulose from pineapple leaf fibres by steam explosion. Carbohydr. Polym. 81: 720–725. https://doi.org/10.1016/j.carbpol.2010.03.046.

Curvello, R., Raghuwanshi, V.S. and Garnier, G. 2019. Engineering nanocellulose hydrogels for biomedical applications. Adv. Colloid. Interface. Sci. https://doi.org/10.1016/j.cis.2019.03.002.

Danial, W.H., Abdul Majid, Z., Mohd Muhid, M.N., Triwahyono, S., Bakar, M.B., Ramli, Z. et al. 2015. The reuse of wastepaper for the extraction of cellulose nanocrystals. Carbohydr. Polym. 118: 165–169. https://doi.org/10.1016/j.carbpol.2014.10.072.

Deepu A., Gopakumar, Daniel Pasquini, H.P.S., Abdul Khalil and Sabu Thomas. 2021. Functional nanocellulose aerogels for varying technological applications. *In*: Nanocellulose. World Scientific (Europe), pp. 449–480. https://doi.org/10.1142/9781786349477_0015.

Fleischmann, S., Mitchell, J.B., Wang, R., Zhan, C., Jiang, D.E., Presser, V. et al. 2020. Pseudocapacitance: From fundamental understanding to high power energy storage materials. Chem. Rev. https://doi.org/10.1021/acs.chemrev.0c00170.

García-González, C.A., Camino-Rey, M.C., Alnaief, M., Zetzl, C. and Smirnova, I. 2012. Supercritical drying of aerogels using CO2: Effect of extraction time on the end material textural properties. Journal of Supercritical Fluids 66: 297–306. https://doi.org/10.1016/j.supflu.2012.02.026.

Geng, S., Wei, J., Jonasson, S., Hedlund, J. and Oksman, K. 2020. Multifunctional carbon aerogels with hierarchical anisotropic structure derived from lignin and cellulose nanofibers for CO2 capture and energy storage. ACS Appl. Mater. Interfaces 12: 7432–7441. https://doi.org/10.1021/acsami.9b19955.

Gu, J., Wang, Z., Kuen, J., Ma, L., Shahroudy, A., Shuai, B. et al. 2018. Recent advances in convolutional neural networks. Pattern. Recognit. 77: 354–377. https://doi.org/10.1016/j.patcog.2017.10.013.

Hassan, M.L., Fadel, S.M., Abouzeid, R.E., Abou Elseoud, W.S., Hassan, E.A., Berglund, L. et al. 2020. Water purification ultrafiltration membranes using nanofibers from unbleached and bleached rice straw. Sci. Rep. 10. https://doi.org/10.1038/s41598-020-67909-3.

Iwamoto, S., Nakagaito, A.N. and Yano, H. 2007. Nano-fibrillation of pulp fibers for the processing of transparent nanocomposites. Appl. Phys. A Mater. Sci. Process 89: 461–466. https://doi.org/10.1007/s00339-007-4175-6.

Jančič, U., Bračič, M., Ojstršek, A., Božič, M., Mohan, T. and Gorgieva, S. 2021. Consolidation of cellulose nanofibrils with lignosulphonate bio-waste into excellent flame retardant and UV blocking membranes. Carbohydr. Polym. 251: 117126. https://doi.org/10.1016/j.carbpol.2020.117126.

Jiménez-Saelices, C., Seantier, B., Cathala, B. and Grohens, Y. 2017. Spray freeze-dried nanofibrillated cellulose aerogels with thermal superinsulating properties. Carbohydr. Polym. 157: 105–113. https://doi.org/10.1016/j.carbpol.2016.09.068.

Kim, J.H., Lee, D., Lee, Y.H., Chen, W. and Lee, S.Y. 2019. Nanocellulose for energy storage systems: Beyond the limits of synthetic materials. Advanced Materials. https://doi.org/10.1002/adma.201804826.

Klemm, D., Kramer, F., Moritz, S., Lindström, T., Ankerfors, M., Gray, D. et al. 2011. Nanocelluloses: A new family of nature-based materials. Angewandte Chemie - International Edition. https://doi.org/10.1002/anie.201001273.

Korhonen, J.T., Kettunen, M., Ras, R.H.A. and Ikkala, O. 2011. Hydrophobic nanocellulose aerogels as floating, sustainable, reusable, and recyclable oil absorbents. ACS Appl. Mater. Interfaces 3: 1813–1816. https://doi.org/10.1021/am200475b.

Lei, C., Xie, Z., Wu, K. and Fu, Q. 2021. Controlled vertically aligned structures in polymer composites: Natural inspiration, structural processing, and functional application. Advanced Materials 33: 2103495. https://doi.org/10.1002/adma.202103495.

Li, V.C.F., Dunn, C.K., Zhang, Z., Deng, Y. and Qi, H.J. 2017. Direct Ink Write (DIW) 3D Printed cellulose nanocrystal aerogel structures. Sci. Rep. 7. https://doi.org/10.1038/s41598-017-07771-y.

Li, Y., Grishkewich, N., Liu, L., Wang, C., Tam, K.C., Liu, S. et al. 2019. Construction of functional cellulose aerogels via atmospheric drying chemically cross-linked and solvent exchanged cellulose nanofibrils. Chemical Engineering Journal 366: 531–538. https://doi.org/10.1016/j.cej.2019.02.111.

Liu, S., Qamar, S.A., Qamar, M., Basharat, K. and Bilal, M. 2021. Engineered nanocellulose-based hydrogels for smart drug delivery applications. Int. J. Biol. Macromol. 181: 275–290. https://doi.org/10.1016/j.ijbiomac.2021.03.147.

Ludwicka, K., Jedrzejczak-Krzepkowska, M., Kubiak, K., Kolodziejczyk, M., Pankiewicz, T. and Bielecki, S. 2016. Medical and cosmetic applications of bacterial nanocellulose. In: Bacterial Nanocellulose: From Biotechnology to Bio-Economy. Elsevier Inc., pp. 145–165. https://doi.org/10.1016/B978-0-444-63458-0.00009-3.

Medina, L., Carosio, F. and Berglund, L.A. 2019. Recyclable nanocomposite foams of Poly(vinyl alcohol), clay and cellulose nanofibrils – Mechanical properties and flame retardancy. Compos. Sci. Technol. 182. https://doi.org/10.1016/j.compscitech.2019.107762.

Mueller, S., Sapkota, J., Nicharat, A., Zimmermann, T., Tingaut, P., Weder, C. et al. 2015. Influence of the nanofiber dimensions on the properties of nanocellulose/poly(vinyl alcohol) aerogels. J. Appl. Polym. Sci. 132. https://doi.org/10.1002/app.41740.

Muthuraj, R., Misra, M. and Mohanty, A.K. 2015. Hydrolytic degradation of biodegradable polyesters under simulated environmental conditions. J. Appl. Polym. Sci. 132. https://doi.org/10.1002/app.42189.

Nechyporchuk, O., Belgacem, M.N. and Bras, J. 2016. Production of cellulose nanofibrils: A review of recent advances. Ind. Crops Prod. https://doi.org/10.1016/j.indcrop.2016.02.016.

Nemoto, J., Saito, T. and Isogai, A. 2015. Simple freeze-drying procedure for producing nanocellulose aerogel-containing, high-performance air filters. ACS Appl. Mater. Interfaces 7: 19809–19815. https://doi.org/10.1021/acsami.5b05841.

Nguyen, S.T., Feng, J., Le, N.T., Le, A.T.T., Hoang, N., Tan, V.B.C. et al. 2013. Cellulose aerogel from paper waste for crude oil spill cleaning. Ind. Eng. Chem. Res. 52: 18386–18391. https://doi.org/10.1021/ie4032567.

Nicu, R., Ciolacu, F. and Ciolacu, D.E. 2021. Advanced functional materials based on nanocellulose for pharmaceutical/medical applications. Pharmaceutics. https://doi.org/10.3390/pharmaceutics13081125.

Norouzi, M., Chàfer, M., Cabeza, L.F., Jiménez, L. and Boer, D. 2021. Circular economy in the building and construction sector: A scientific evolution analysis. Journal of Building Engineering 44: 102704. https://doi.org/10.1016/j.jobe.2021.102704.

Pääkko, M., Ankerfors, M., Kosonen, H., Nykänen, A., Ahola, S., Österberg, M. et al. 2007. Enzymatic hydrolysis combined with mechanical shearing and high-pressure homogenization for nanoscale cellulose fibrils and strong gels. Biomacromolecules 8: 1934–1941. https://doi.org/10.1021/bm061215p.

Pelissari, F.M., Sobral, P.J.D.A. and Menegalli, F.C. 2014. Isolation and characterization of cellulose nanofibers from banana peels. Cellulose 21: 417–432. https://doi.org/10.1007/s10570-013-0138-6.

Plappert, S.F., Quraishi, S., Nedelec, J.M., Konnerth, J., Rennhofer, H., Lichtenegger, H.C. et al. 2018. Conformal ultrathin coating by scCO2-mediated PMMA deposition: A facile approach to add moisture resistance to lightweight ordered nanocellulose aerogels. Chemistry of Materials 30: 2322–2330. https://doi.org/10.1021/acs.chemmater.7b05226.

Pottathara, Y.B., Bobnar, V., Finšgar, M., Grohens, Y., Thomas, S., Kokol, V. et al. 2018. Cellulose nanofibrils-reduced graphene oxide xerogels and cryogels for dielectric and electrochemical storage applications. Polymer (Guildf) 147: 260–270. https://doi.org/10.1016/j.polymer.2018.06.005.

Qu, R., Zhang, W., Liu, N., Zhang, Q., Liu, Y., Li, X. et al. 2018. Antioil Ag_3PO_4 Nanoparticle/ Polydopamine/Al_2O_3 sandwich structure for complex wastewater treatment: Dynamic catalysis under natural light. ACS Sustain. Chem. Eng. 6: 8019–8028. https://doi.org/10.1021/acssuschemeng.8b01469.

Sheikhy, S., Safekordi, A.A., Ghorbani, M., Adibkia, K. and Hamishehkar, H. 2021. Synthesis of novel superdisintegrants for pharmaceutical tableting based on functionalized nanocellulose hydrogels. Int. J. Biol. Macromol. 167: 667–675. https://doi.org/10.1016/j.ijbiomac.2020.11.173.

Subhedar, A., Bhadauria, S., Ahankari, S. and Kargarzadeh, H. 2021. Nanocellulose in biomedical and biosensing applications: A review. Int. J. Biol. Macromol. https://doi.org/10.1016/j.ijbiomac.2020.10.217.

Sun, F., Liu, W., Dong, Z. and Deng, Y. 2017. Underwater superoleophobicity cellulose nanofibril aerogel through regioselective sulfonation for oil/water separation. Chemical Engineering Journal 330: 774–782. https://doi.org/10.1016/j.cej.2017.07.142.

Tang, J., He, H., Wan, R., Yang, Q., Luo, H., Li, L. et al. 2021. Cellulose nanocrystals for skin barrier protection by preparing a versatile foundation liquid. ACS Omega 6: 2906–2915. https://doi.org/10.1021/acsomega.0c05257.

Toivonen, M.S., Kaskela, A., Rojas, O.J., Kauppinen, E.I. and Ikkala, O. 2015. Ambient-dried cellulose nanofibril aerogel membranes with high tensile strength and their use for aerosol collection and templates for transparent, flexible devices. Adv. Funct. Mater. 25: 6618–6626. https://doi.org/10.1002/adfm.201502566.

Tummala, G.K., Joffre, T., Rojas, R., Persson, C. and Mihranyan, A. 2017. Strain-induced stiffening of nanocellulosereinforced poly(vinyl alcohol) hydrogels mimicking collagenous soft tissues. Soft Matter 13: 3936–3945. https://doi.org/10.1039/c7sm00677b.

Yang, J., Xia, Y., Xu, P. and Chen, B. 2018. Super-elastic and highly hydrophobic/ superoleophilic sodium alginate/cellulose aerogel for oil/water separation. Cellulose 25: 3533–3544. https://doi.org/10.1007/s10570-018-1801-8.

Yassin, S., Goodwin, D.J., Anderson, A., Sibik, J., Wilson, D.I., Gladden, L.F. et al. 2015. The disintegration process in microcrystalline cellulose based tablets, Part 1: Influence of temperature, porosity and superdisintegrants. J. Pharm. Sci. 104: 3440–3450. https://doi.org/10.1002/jps.24544.

Zeng, Z., Wu, T., Han, D., Ren, Q., Siqueira, G., Nyström, G. et al. 2020. Ultralight, flexible, and biomimetic nanocellulose/silver nanowire aerogels for electromagnetic interference shielding. ACS Nano 14: 2927–2938. https://doi.org/10.1021/acsnano.9b07452.

Zhang, X., Elsayed, I., Navarathna, C., Schueneman, G.T. and Hassan, E.B. 2019. Biohybrid hydrogel and aerogel from self-assembled nanocellulose and nanochitin as a high-efficiency adsorbent for water purification. ACS Appl. Mater. Interfaces 11: 46714–46725. https://doi.org/10.1021/acsami.9b15139.

Zhang, X., Zhao, X., Xue, T., Yang, F., Fan, W., Liu, T. et al. 2020. Bidirectional anisotropic polyimide/bacterial cellulose aerogels by freeze-drying for super-thermal insulation. Chemical Engineering Journal 385. https://doi.org/10.1016/j.cej.2019.123963.

Zu, G., Shen, J., Zou, L., Wang, F., Wang, X., Zhang, Y. et al. 2016. Nanocellulose-derived highly porous carbon aerogels for supercapacitors. Carbon N Y 99: 203–211. https://doi.org/10.1016/j.carbon.2015.11.079.

CHAPTER 3

Nanomaterials
Future of Light-driven and UV-photodetector Materials

Muhammad Qasim

1. Introduction

Nanomaterials are materials that have at least one dimension in the nanoscale range, typically between 1 to 100 nanometers (Santos et al. 2015). At this scale, materials exhibit unique physical and chemical properties that can differ from their bulk counterparts (Yamashita and Li 2016). In industrial chemistry, nanomaterials have many potential applications. Some of these include: (1) catalysis: nanoparticles are often used as catalysts because of their large surface area-to-volume ratio, which makes them highly effective in facilitating chemical reactions; by using nanoparticles as catalysts, chemical reactions can occur faster, with less energy input and with fewer unwanted byproducts, thus making chemical processes more efficient and environmentally friendly; (2) coatings: nanomaterials can be used as coatings to improve the properties of surfaces; for example, adding nanoparticles to a coating can make it more resistant to scratches, wear or UV light; also, nano coatings can be designed to have self-cleaning properties, making them useful in applications such as water-repellent coatings or anti-fogging surfaces; (3) energy storage: nanomaterials can be used to improve the performance of energy storage devices, such as batteries or capacitors; by using nanoparticles as electrodes, the surface area of the electrode is increased, allowing for faster charging and

Department of Civil Engineering, The University of Lahore, Lahore Campus, 1 KM-Defense Road, Lahore, Pakistan.
Email: muhammad.qasim@ce.uol.edu.pk

discharging of the device; nanomaterials can also be used to increase the energy density of the device, implying that it can store more energy in a smaller volume; (4) sensors: nanomaterials are highly sensitive to changes in their environment, making them ideal for use in sensors; for instance, nanoparticles can be used to detect and measure the presence of a wide range of chemicals and biological molecules. Additionally, the small size of nanoparticles makes them useful for sensing at nanoscale, such as in single-molecule detection or in biosensors for medical diagnostics; (5) drug delivery: nanomaterials can be designed to act as carriers for drugs and other therapeutics, enabling targeted delivery to specific areas of the body. By encapsulating drugs in nanoparticles, they can be protected from degradation and can be delivered more effectively to their target site (Wang et al. 2021a). Additionally, nanoparticles can be designed to release the drug slowly over time, allowing for sustained release and prolonged therapeutic effect (Wang et al. 2019).

Light-driven nanomaterials are very promising materials as they have the potential to revolutionize many areas of science and technology (Ansari et al. 2021). By using light as a means of triggering chemical reactions or controlling the properties of materials, light-driven nanomaterials can enable new applications in fields such as medicine, energy and electronics (Hussain 2020). One promising area of research is the development of light-driven nanomachines, which are tiny devices that can be controlled by using light to perform specific tasks. These devices can be used for targeted drug delivery, sensing and monitoring, or even for performing complex computations at the nanoscale (Hussain 2020). Another exciting application of light-driven nanomaterials is in the field of photonics (the study of light and its interactions with matter). By using light to control the properties of nanomaterials, researchers can create new types of optical materials with unique properties that could be used in applications such as telecommunications, data storage and imaging (Karthikeyan et al. 2020).

Photodetectors are nanoscale electronic devices that can detect and convert light into electrical signals. They are widely used in many applications such as telecommunications, photography, spectroscopy and sensing (Donati and Technology 2001). The basic principle of a photodetector is based on the photoelectric effect—this refers to the ejection of electrons from a material when photons of light are incident on it. There are several types of photodetectors, including photodiodes, phototransistors, photoconductors and photomultipliers. Photodiodes are the most common type of photodetectors and consist of a p-n junction that produces a current when illuminated by light (Donati and Technology 2001). The amount of current produced as a result is proportional to the amount of light incident on the photodiode. Phototransistors, on the other hand, are like photodiodes, but with the added amplification of a transistor.

When light strikes the junction, they produce a current that is amplified by the transistor, resulting in a larger output signal. Photoconductors are another type of photodetector; they operate on the principle of change in conductivity due to the absorption of light (Long et al. 2019). When light is incident on a photoconductive material, it creates more free electrons, increasing the conductivity of the material and in turn resulting in a change in output voltage or current. Photomultipliers are a highly sensitive type of photodetector that can amplify very low levels of light. They consist of a vacuum tube with a photocathode that produces electrons when exposed to light (Chen et al. 2015). These electrons are then multiplied through a series of dynodes, resulting in a large output signal. Photodetectors are essential devices that are used to detect and measure light in a wide range of applications. The choice of a particular type of photodetector depends on the requirements of the application, such as sensitivity, response time and spectral range (Donati and Technology 2001).

Light-driven nanomaterials, such as quantum dots and nanowires, have unique optical properties that make them promising candidates for use in photodetectors. A photodetector is an essential component of many modern technologies, including digital cameras, solar cells and medical imaging systems (Long et al. 2019). The efficiency of this process depends on the properties of the material used. Light-driven nanomaterials have a high surface-area-to-volume ratio, which means that they can absorb more light per unit of material, compared to bulk materials. Additionally, the electronic properties of nanomaterials can be tuned by controlling their size and composition, allowing for precise tuning of the absorption spectrum (Wu et al. 2015). Quantum dots are one type of light-driven nanomaterial that have been extensively studied for use in photodetectors. Quantum dots are small semiconductor particles with dimensions in the order of nanometers (Ali et al. 2022). Due to their small size, they exhibit quantum confinement effects, resulting in discrete energy levels and tunable optical properties. By controlling the size and composition of quantum dots, their absorption spectrum can be tuned to match the wavelength of the incident light, thus maximizing their efficiency as photodetectors (Riaz et al. 2019). Nanowires are yet another type of light-driven nanomaterial that have been investigated for use in photodetectors. Nanowires are elongated structures with a diameter of a few nanometers and length of several micrometers (Yamashita and Li 2016). They have a large surface area, which makes them highly efficient at absorbing light. Additionally, the electronic properties of nanowires can be tuned by controlling their diameter and composition, allowing for the precise tuning of their absorption spectrum. In conclusion, light-driven nanomaterials have the potential to revolutionize the field of photodetectors by offering higher efficiencies and greater control over the absorption spectrum.

Quantum dots and nanowires are two examples of light-driven nanomaterials that have shown promise for use in photodetectors.

2. Photocatalysis

Photocatalysis is a process in which light is used to activate a catalyst which can subsequently drive a chemical reaction. The catalyst used in this process is typically a semiconductor material such as titanium dioxide (TiO_2) or zinc oxide (ZnO) (Abdullah et al. 2022, Gautam et al. 2020). When light hits the surface of the catalyst, it creates electron-hole pairs, which can then participate in chemical reactions, as exhibited in Figure 1. The photocatalysis process can be divided into several steps:

1. Absorption of light: When light of a specific wavelength hits the surface of the photocatalytic material, it is absorbed by the electrons in the material, leading to the creation of electron-hole pairs. The energy required for this process is typically supplied by UV light.

2. Separation of electron-hole pairs: Once the electron-hole pairs are created, they must be separated before they can participate in chemical reactions. This is typically achieved through the use of an electric field within the material or by physically separating the electron and hole regions.

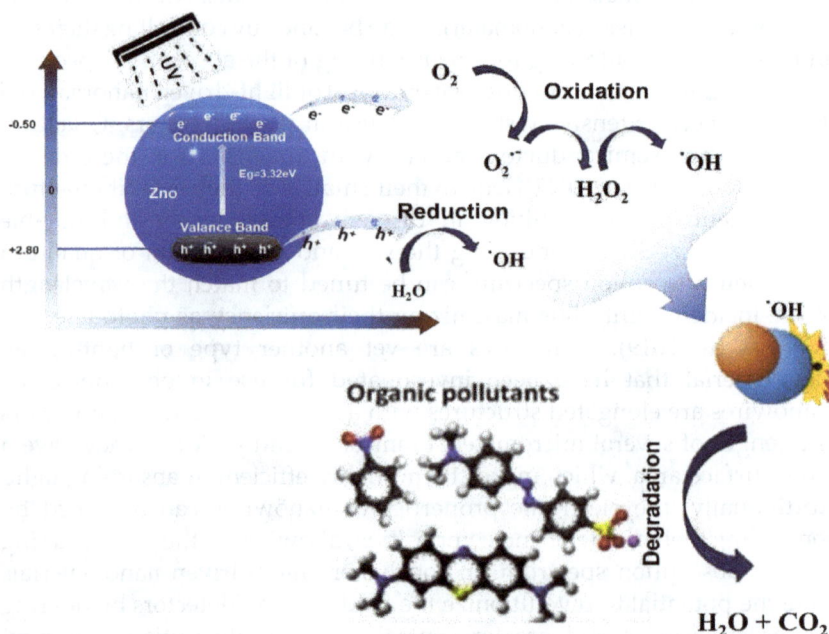

Figure 1: Photocatalytic degradation of pollutants using ZnO nanoparticles.

3. Formation of reactive species: The electron-hole pairs that are separated can interact with molecules in the surrounding environment to create reactive species such as hydroxyl radicals ($OH^{\bullet}$) or superoxide radicals ($O_2^{\bullet-}$). These species can then participate in chemical reactions with other molecules in the environment.

4. Chemical reactions: Once the reactive species are formed, they can react with organic pollutants or other molecules in the environment, leading to their degradation or transformation into other compounds.

The overall efficiency of the photocatalysis process depends on several factors, including the properties of the photocatalyst, the intensity and wavelength of the light source, and the chemical properties of the target molecules (Hou et al. 2014). With the help of nanomaterials, photocatalysis has become an increasingly important process with a wide range of applications, including environmental remediation, energy production, and chemical synthesis.

In the context of industrial applications, photocatalysis can be used to promote various chemical reactions for a wide range of products. One of the main advantages of photocatalysis is that it can be performed at ambient temperature and pressure, which makes it a more energy-efficient and sustainable process compared to traditional chemical reactions (Perović et al. 2020). Nanomaterials are particularly useful in photocatalysis because they have a high surface-area-to-volume ratio, which allows for more efficient interaction with light and catalytic reactions. Some common examples of nanomaterials used in photocatalysis include titanium dioxide (TiO_2), zinc oxide (ZnO) and graphene oxide (GO) (Abdullah et al. 2022, Karthik et al. 2021, Mishra et al. 2022, Perović et al. 2020, Wu et al. 2015, Yang et al. 2020). One major industrial application of photocatalysis is in the field of water treatment. Nanomaterial-based photocatalysts can be used to remove contaminants such as organic pollutants, heavy metals and bacteria from wastewater as well as drinking water (Wang et al. 2021a). For example, TiO_2 nanoparticles have been used to decompose organic pollutants in water by breaking down their chemical bonds under UV light. Similarly, ZnO nanoparticles have been used to kill bacteria in water by generating reactive oxygen species (ROS) that damagetheir cellular membranes (Santos et al. 2015). Another important application of photocatalysis is in the field of energy conversion (Santos et al. 2015). Nanomaterial-based photocatalysts can be used to split water into hydrogen and oxygen, using sunlight as the energy source (Santos et al. 2015). This process, called photocatalytic water splitting, has the potential to provide a renewable source of hydrogen fuel for various industrial applications (Perović et al. 2020). Similarly, nanomaterial-based photocatalysts can be used to convert carbon dioxide (CO_2) into fuels such as methane, methanol and ethanol, with sunlight as the energy source (Nahar et al. 2017). This process, called photocatalytic CO_2 reduction, has the potential to reduce greenhouse gas emissions and provide a renewable source of fuels (Tran et al. 2022). In addition to these applications, photocatalysis has potential uses in other industrial applications such as air purification, self-cleaning surfaces, and medical applications.

3. Classification of Light-driven Nanomaterials

Light-driven nanomaterials are a type of nanomaterial that can respond to light in different ways, including generation of electricity, initiating chemical reactions, changing their color or properties, and controlling the activity of cells (Hou et al. 2014). These materials are typically made up of nanoparticles or nanocomposites that have unique optical and electronic properties which enable them to interact with light in specific ways (Karthikeyan et al. 2020). In the broader aspect, light-driven nanomaterials can be classified into several categories based on their functionality and mechanisms of operation (Abdullah et al. 2022). Here are some of the common classifications:

1. Photocatalysts: These materials absorb light energy and use it to initiate a chemical reaction, usually by generating electron-hole pairs in the material. Examples in this category include titanium dioxide (TiO_2) and zinc oxide (ZnO), which are commonly used in environmental remediation, such as wastewater treatment and air purification.

2. Photochromic materials: These materials change their color or optical properties in response to light. An example of such a material is spiropyran, a molecule that undergoes a reversible transformation between a colorless, non-fluorescent form and a colored, fluorescent form when exposed to UV light.

3. Photovoltaic materials: These materials generate electricity through the conversion of light energy into electrical energy. Common examples herein include silicon-based solar cells and organic photovoltaics (OPVs), which are lightweight, flexible and can be produced using low-cost printing techniques.

4. Photoelectrochemical materials: These materials use light energy to drive a chemical reaction that produces electrical energy. For instance, a dye-sensitized solar cell (DSSC) is a photoelectrochemical material and consists of a photosensitizer and a semiconductor layer that absorbs light and generates electricity.

5. Light-responsive biomaterials: These materials respond to light by changing their shape or properties. For example, photoresponsive hydrogels are light-responsive biomaterials which can be used for

drug delivery or tissue engineering, and optogenetic tools, allowing scientists to control the activity of cells using light.

The industrial utilization of light-driven nanomaterials can be classified based on their applications in various fields such as environmental remediation, energy production, and chemical synthesis (Belousov and Suleimanov 2021, Hussain 2020, Lopez et al. 2021). Some of them are listed as follows:

1. Environmental remediation: Light-driven nanomaterials can be used for the degradation of organic pollutants in air and water. For example, photocatalysts such as TiO_2 and ZnO can be used to degrade organic pollutants in wastewater and carbon-based nanomaterials can be used to remove heavy metals and other contaminants from water.

2. Energy production: Light-driven nanomaterials can be used for the production of clean energy, such as hydrogen fuel. For example, metal oxide nanomaterials such as TiO_2 and ZnO can be used for photocatalytic water splitting to produce hydrogen fuel. In addition, nanomaterials such as perovskites and quantum dots can be used to improve the efficiency of solar cells.

3. Chemical synthesis: Light-driven nanomaterials can be used for the synthesis of chemicals and materials. For example, metal-organic frameworks (MOFs) and other porous materials can be synthesized using light-driven processes. Nanocatalysts can also be used for the selective synthesis of chemicals, selective hydrogenation of alkenes for instance.

4. Biomedical applications: Light-driven nanomaterials can be used for biomedical applications such as drug delivery and imaging. For example, gold and iron oxide nanoparticles can be used for photothermal therapy and magnetic resonance imaging (MRI) respectively. In addition, light-driven nanomaterials can be used for targeted drug delivery using photochemical activation.

Overall, light-driven nanomaterials have a wide range of applications in various fields and can offer unique advantages such as high efficiency, selectivity and specificity.

4. Light Harvesting and Optical Characteristics of Photocatalytic Nanomaterials

The light harvesting characteristics of these materials depend on a number of factors, including material composition, size, shape and surface properties. One important factor that affects the light harvesting efficiency of photocatalytic nanomaterials is their band gap. The band gap determines the wavelength of light that can be absorbed by the material. Materials with a smaller band gap are capable of absorbing a broader range of wavelengths and therefore have higher light harvesting efficiency (Kandy and Fuels 2020). The size and shape of nanomaterials also play a crucial role in determining their light harvesting characteristics. Small-sized nanoparticles have a larger surface-area-to-volume ratio, which increases their light harvesting efficiency (Mishra et al. 2022). Additionally, certain shapes such as nanorods and nanowires have been found to enhance the light harvesting properties of nanomaterials due to their unique optical properties. The surface properties of photocatalytic nanomaterials also affect their light harvesting efficiency. The presence of defects or impurities on the surface of the nanomaterials can lead to a decrease in their light harvesting efficiency by trapping the generated charge carriers (Wang et al. 2021a). On the other hand, surface modifications such as doping or functionalization can enhance the light harvesting efficiency by improving the charge transfer and separation processes.

The optical characteristics of photocatalytic nanomaterials are important for their photocatalytic activity, which involves the absorption of light energy to initiate chemical reactions (Lettieri et al. 2021). Here are some of the key optical characteristics of photocatalytic nanomaterials:

1. Absorption spectrum: The absorption spectrum of a photocatalytic nanomaterial is the range of wavelengths of light that it can absorb. The absorption spectrum is determined by the band gap of the material, which is the minimum energy required to excite an electron from the valence band to the conduction band. Materials with a smaller band gap can absorb a broader range of wavelengths and have higher light harvesting efficiencies.

2. Reflectance and transmittance: The reflectance and transmittance of a photocatalytic nanomaterial describe how much light is reflected or transmitted through the material, respectively. These properties can be important for designing photocatalytic devices that require light to be transmitted through the material to reach the reaction site.

3. Photoluminescence: Photoluminescence is the emission of light by a material upon excitation by a light source. Photocatalytic nanomaterials can exhibit photoluminescence when excited by light, which can be useful for studying the materials' electronic properties and understanding their photocatalytic activity.

4. Surface plasmon resonance: Some photocatalytic nanomaterials can exhibit surface plasmon resonance, a phenomenon that occurs when light interacts with the free electrons on the surface of a metal nanoparticle. Surface plasmon resonance can enhance the light

absorption and scattering properties of the material, in turn improving its photocatalytic activity.

5. Quantum confinement effects: When the size of a nanomaterial is reduced below a certain threshold, quantum confinement effects can occur. These effects arise from the discrete energy levels of the nanomaterial, which can shift the absorption and emission spectra of the material to shorter wavelengths. This can enhance the photocatalytic activity of the material by increasing the number of high-energy photons that can be absorbed.

Overall, the optical characteristics of photocatalytic nanomaterials are closely related to their photocatalytic activity and can be tuned by controlling the material composition, size and shape. Understanding these properties is essential for designing and optimizing photocatalytic materials for various applications.

5. Synthesis Techniques of Photocatalytic Nanomaterials

The synthesis of photocatalytic nanomaterials typically involves the preparation of nanoparticles with a high surface-area-to-volume ratio, allowing for increased photocatalytic activity (Abdullah et al. 2022, Akhoondi et al. 2021, Kolahalam et al. 2019, Sakdaronnarong et al. 2020, Wang et al. 2021b, Yadav and Jaiswar 2017). Here are some common methods of synthesis of photocatalytic nanomaterials:

1. Sol-gel method: This method involves the formation of a gel-like precursor solution that is then transformed into solid nanomaterials via drying and calcination. This method is versatile and can be used to synthesize a wide range of materials.

2. Hydrothermal method: This method involves the use of high-pressure and high-temperature conditions to synthesize nanomaterials. This method is particularly useful for the synthesis of metal oxides and sulfides.

3. Solvothermal method: This method is similar to the hydrothermal method but involves the use of a solvent to control the size and shape of nanoparticles.

4. Chemical vapor deposition (CVD): This method involves the deposition of precursor gases onto a substrate which subsequently undergoes a chemical reaction to form the desired nanomaterial. This method is useful for the synthesis of thin films.

5. Microwave-assisted synthesis: This method uses microwaves to heat the precursor solution, resulting in the rapid and efficient synthesis of nanomaterials.

6. Electrochemical deposition: This method uses an electrochemical cell to deposit nanomaterials on to a substrate. This method is particularly useful for the synthesis of metal nanoparticles.

These methods can be tailored to produce a wide range of photocatalytic nanomaterials with varying properties, making them useful for a variety of applications such as water purification, air purification and renewable energy production.

6. Immobilization, Surface Modification and Shape Engineering of Photocatalytic Nanomaterials

Photocatalytic nanomaterials have gained significant attention in recent years due to their potential applications in various fields such as water purification, air purification and self-cleaning surfaces (Bui et al. 2021, Mamba and Mishra 2016, Nahar et al. 2017, Wang et al. 2021a). The tunability of nanomaterials leading to various modifications in size, shape, surface charge, aggregation and suspension potential (exhibited in Figure 2) leads to their vast application in various fields and industrial processes (Chen et al. 2020, Görke and Garnweitner 2021, Hao et al. 2017, Valizadeh et al. 2018).

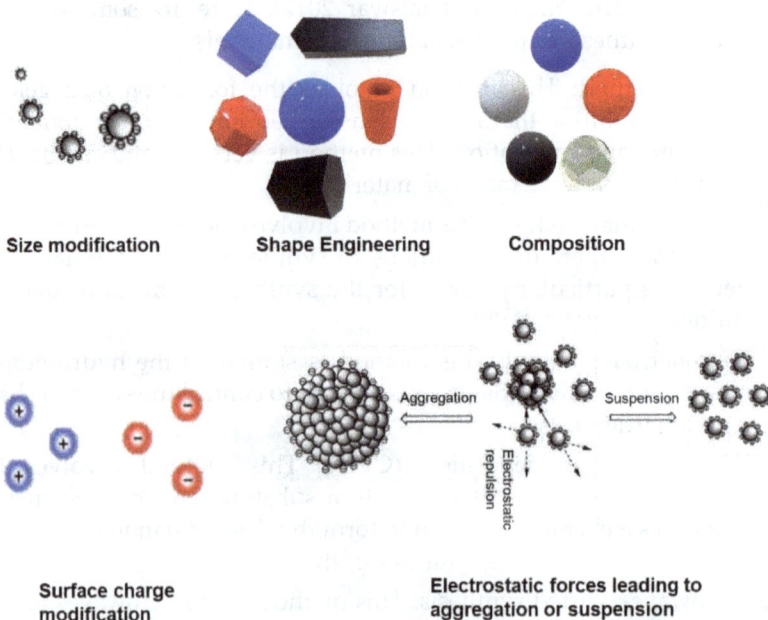

Figure 2: Tunable properties of photocatalytic nanomaterials corresponding to application requirements.

The immobilization of these nanomaterials is crucial to prevent their leaching, aggregation and loss of activity. Here are some common immobilization techniques for photocatalytic nanomaterials:

1. Deposition on solid substrates: Photocatalytic nanomaterials can be deposited on solid substrates such as glass, ceramics, or metal plates. This method is suitable for large-scale applications and ensures a stable and durable photocatalytic system.

2. Encapsulation in polymer matrices: Photocatalytic nanomaterials can be encapsulated in polymer matrices such as polyvinyl alcohol, polyethylene glycol and polystyrene. This method protects nanomaterials from leaching, aggregation and environmental factors, and enhances their stability.

3. Immobilization on fibers: Photocatalytic nanomaterials can be immobilized on fibers such as cotton, wool and polyester. This method is suitable for air purification applications and provides a large surface area for the photocatalytic reaction.

4. Surface Coatings Photocatalytic nanomaterials can be coated on surfaces such as walls, windows or tiles; this method is suitable for self-cleaning surfaces and enhances the photocatalytic activity of these surfaces.

5. Integration with membranes: The photocatalytic nanomaterials can be integrated with membranes such as ultrafiltration, nanofiltration or reverse osmosis membranes. This method is suitable for water purification applications and enhances the removal of organic pollutants and bacteria.

Overall, the immobilization of photocatalytic nanomaterials is crucial for enhancing their stability, durability, reusability and activity in various applications (Cipolatti et al. 2016; Putzbach and Ronkainen 2013, Shuai et al. 2017). Immobilization of photocatalytic nanomaterials is performed through the following techniques:

1. Sol-gel method: In this technique, photocatalytic nanomaterials are mixed with a sol-gel precursor, such as tetraethyl orthosilicate, to form a homogeneous solution. The mixture is then deposited on to a substrate and allowed to gel, forming a stable immobilized photocatalyst.

2. Electrospinning: In this technique, photocatalytic nanomaterials are dispersed in a polymer solution and electro spun onto a substrate. The resulting nanofibers provide a high surface area for photocatalytic reactions and also enhance the mechanical stability of the immobilized photocatalyst.

3. Co-precipitation: In this technique, photocatalytic nanomaterials are co-precipitated with a support material, such as silica or alumina, to form a stable immobilized photocatalyst. The co-precipitation process allows for the formation of a homogeneous mixture, ensuring an even distribution of photocatalytic nanomaterials on the support material.

4. Dip coating: In this technique, a substrate is dipped into a solution containing photocatalytic nanomaterials, and the excess solution is removed by drying or rinsing. The process is repeated multiple times to form a stable and uniform layer of immobilized photocatalyst on the substrate.

5. Spray coating: In this method, photocatalytic nanomaterials are dispersed in a solution and sprayed on to a substrate using a spray gun. The resulting layer of immobilized photocatalyst is thin and uniform, allowing for efficient photocatalytic reactions.

Ultimately, the choice of immobilization technique depends on the properties of the photocatalytic nanomaterial, the substrate, and the desired application.

Surface modification of photocatalytic nanomaterials is an essential process that can improve their properties and enhance their performance in various applications (Wen et al. 2015, Zhao et al. 2022, Zheng et al. 2018). Some common surface modification techniques for photocatalytic nanomaterials include:

1. Doping: Doping involves introducing impurities or foreign atoms into photocatalytic nanomaterials to modify their electronic properties. For example, doping titania (TiO_2) with nitrogen (N) can shift the absorption edge of TiO_2 to the visible light region, thus enhancing its photocatalytic activity under visible light. Similarly, doping with other impurities or foreign atoms such as sulfur (S), carbon (C), or metal ions can also modify the electronic properties of photocatalytic nanomaterials and improve their photocatalytic activity.

2. Coating: Coating involves depositing a thin layer of a secondary material onto the surface of photocatalytic nanomaterials to modify their properties. For example, coating TiO_2 nanoparticles with a layer of silicon dioxide (SiO_2) can increase their stability and reduce their aggregation, leading to improved photocatalytic activity. Coating photocatalytic nanomaterials with a layer of metal oxide, such as zinc oxide (ZnO), can also improve their stability and increase their light absorption capacity.

3. Functionalization: Functionalization involves attaching organic or inorganic molecules to the surface of photocatalytic nanomaterials to modify their surface chemistry and properties. For example, functionalizing the surface of TiO_2 nanoparticles with carboxylic acids

can improve their dispersibility and stability in aqueous solutions. Similarly, functionalizing the surface of photocatalytic nanomaterials with amino acids, thiol groups, or other functional groups can also modify their surface chemistry and improve their photocatalytic activity.

4. Sensitization: Sensitization involves attaching a light-absorbing molecule to the surface of photocatalytic nanomaterials to enhance their light absorption capacity. For example, sensitizing the surface of TiO_2 nanoparticles with organic dyes or metal complexes can enhance their photocatalytic activity under visible light. Sensitization can also improve the selectivity of photocatalytic reactions by enhancing the absorption of specific wavelengths of light.

5. Surface texturing: Surface texturing involves modifying the surface morphology of photocatalytic nanomaterials to increase their surface area and improve their light absorption capacity. For example, creating a rough surface on the surface of TiO_2 nanoparticles can increase their surface area and enhance their photocatalytic activity. Surface texturing can also enhance the charge separation and transfer processes, leading to improved photocatalytic activity.

Overall, surface modification techniques can improve the photocatalytic activity, selectivity and stability of photocatalytic nanomaterials and expand their range of applications. The choice of surface modification technique(s) depends on the properties of the photocatalytic nanomaterials, the desired modifications and the intended applications (Wen et al. 2015). There are several methods of surface modification of photocatalytic nanomaterials, including:

1. Sol-gel method: The sol-gel method involves the hydrolysis and condensation of metal alkoxides to form a metal oxide coating on the surface of photocatalytic nanomaterials. This method can be used to deposit a thin, uniform layer of metal oxide on photocatalytic nanomaterials to improve their stability and enhance their photocatalytic activity.

2. Chemical vapor deposition (CVD): CVD involves the deposition of a thin film of a material on the surface of photocatalytic nanomaterials using a gas-phase reaction. This method can be used to deposit a variety of materials, including metal oxides, on photocatalytic nanomaterials in order to modify their properties.

3. Electrochemical deposition: Electrochemical deposition uses an electric current to deposit a layer of a material on the surface of photocatalytic nanomaterials. This method can be used to deposit a thin, uniform layer of metal or metal oxide on to photocatalytic nanomaterials to modify their properties.

4. Plasma treatment: Plasma treatment involves the use of a low-temperature plasma to modify the surface chemistry and properties of photocatalytic nanomaterials. This method can be used to functionalize the surface of photocatalytic nanomaterials by introducing functional groups, or to modify the morphology of the surface to enhance its photocatalytic activity.

5. Photodeposition: Photodeposition uses light to deposit a layerof a material on the surface of photocatalytic nanomaterials. This method can be used to deposit a variety of materials, including metals and metal oxides, onto photocatalytic nanomaterials to modify their properties.

6. Chemical modification: Chemical modification involves the use of chemical reactions to modify the surface chemistry and properties of photocatalytic nanomaterials. This method can be used to introduce functional groups on to the surface of photocatalytic nanomaterials, or to modify the morphology of the surface to enhance its photocatalytic activity.

Overall, the choice of surface modification method depends on the properties of the photocatalytic nanomaterials, the desired modifications and the intended applications. Each of these methods has its own advantages and disadvantages and should be selected based on the specific requirements of the application.

Shape engineering of photocatalytic nanomaterials is the process of controlling the morphology and shape of nanoparticles to improve their photocatalytic activity and selectivity. The morphology of nanoparticles plays an important role in their photocatalytic performance, as it affects the surface area, the number and distribution of active sites, charge transfer, and the light absorption properties of these nanoparticles (Valizadeh et al. 2018). Therefore, controlling the shape and size of nanoparticles is an important factor in optimizing their photocatalytic activity for specific applications. Several methods can be used for the shape engineering of photocatalytic nanomaterials, including template-assisted synthesis, hydrothermal synthesis, solvothermal synthesis, seed-mediated growth and electrochemical synthesis (Chen et al. 2020, Hao et al. 2017, Valizadeh et al. 2018). These methods, as explained below, can be used to control the growth and shape of nanoparticles by adjusting various reaction parameters such as temperature, pressure, solvent, and current density.

1. Template-assisted synthesis involves a template to control the shape and size of nanoparticles. The template can be a surfactant, a polymer or a biological molecule. This method can be used to produce nanoparticles with unique shapes and sizes, such as hexagonal-prism-shaped TiO_2 nanoparticles.

2. Hydrothermal synthesis and solvothermal synthesis are similar methods that involve the heating of a solution of precursor chemicals to induce the growth of nanoparticles with specific shapes and sizes. The morphology of the nanoparticles can be controlled by adjusting the reaction conditions such as temperature, pressure and reaction time. For example, hydrothermal synthesis of TiO_2 at high temperature and pressure can result in the formation of one-dimensional nanorods.

3. Seed-mediated growth involves the use of small precursor nanoparticles as seeds to control the growth of larger nanoparticles with specific shapes and sizes. This method can be used to produce nanoparticles with different shapes, such as nanorods or nanotubes, using silver nanoparticles as seeds for the growth of TiO_2 nanoparticles.

4. Electrochemical synthesis uses an electric current to control the growth of nanoparticles on an electrode surface. The morphology of the nanoparticles can be controlled by adjusting the current density and reaction time. This method can be used to produce nanoparticles with unique shapes such as cubic or octahedral shapes.

To summarize, shape engineering of photocatalytic nanomaterials can be achieved by using various methods that control their morphology and shape. These methods can improve the photocatalytic activity and selectivity of nanoparticles by optimizing their surface area, charge transfer, and light absorption properties for specific applications.

7. Applications in Solar Light Harvesting and Energy Storage

Recent research in photocatalytic nanomaterials has focused on improving their efficiency of solar light harvesting for various applications (Gautam et al. 2020, Kandy and Fuels 2020, Karthikeyan et al. 2020, Saber et al. 2020, Yang et al. 2020). Some of the recent research trends and successful applications of photocatalytic nanomaterials in solar light harvesting are as follows:

1. Visible-light-responsive photocatalysts: Most photocatalytic nanomaterials such as TiO_2 are only activated by UV light, which limits their applications in solar light harvesting. Recent research has focused on developing visible-light-responsive photocatalysts that can harvest a broader range of solar radiation, such as carbon nitride, graphitic carbon nitride, and metal-organic frameworks (MOFs).

2. Heterojunctions and co-catalysts: Another research trend uses heterojunctions and co-catalysts to enhance the photocatalytic activity of nanomaterials. Heterojunctions involve combining two

different types of nanomaterials with complementary band structures to improve the separation and transfer of photo-generated charges. Co-catalysts, on the other hand, are metallic or non-metallic materials that act as electron acceptors or donors to enhance the photocatalytic activity of nanomaterials.

3. Water splitting and hydrogen production: Photocatalytic nanomaterials have also been successfully applied in the field of water splitting and hydrogen production. These materials can harvest solar radiation to generate hydrogen fuel from water. For example, TiO_2 nanotubes and graphitic carbon nitride have been used to efficiently split water into hydrogen and oxygen under visible light radiation.

4. Photovoltaics: Photocatalytic nanomaterials have also been applied in photovoltaic devices such as dye-sensitized solar cells (DSSCs) and perovskite solar cells. For example, TiO_2 nanoparticles are commonly used as the electron transport layer in DSSCs, given their high electron mobility and good chemical stability.

It can be seen that recent research trends in photocatalytic nanomaterials have focused on improving their efficiency in solar light harvesting for various applications such as water splitting, environmental remediation, and photovoltaics. These efforts have led to the successful application of photocatalytic nanomaterials in various fields, highlighting their potential as a sustainable and clean energy source. Moreover, photocatalytic nanomaterials have been extensively studied for their potential applications in energy storage. Specifically, photocatalytic nanomaterials can be used to harvest solar energy and convert it into chemical energy which can be stored in the form of fuels such as hydrogen or hydrocarbons (Wang et al. 2019). Some of the common photocatalytic nanomaterials used for energy storage applications include TiO_2, ZnO, CdS and GaN (Karthikeyan et al. 2020).

There are two primary ways in which photocatalytic nanomaterials can be used for energy storage:

1. Hydrogen production: Photocatalytic nanomaterials can be used to produce hydrogen gas from water, using sunlight. This process, called photocatalytic water splitting, uses the energy from the sun to excite electrons in the nanomaterial, which can then be used to split water molecules into hydrogen and oxygen. Hydrogen gas produced in this manner can be stored and used as a fuel for various applications such as powering vehicles or generating electricity.

2. Carbon dioxide reduction: Photocatalytic nanomaterials can also be used to reduce carbon dioxide (CO_2) into hydrocarbons which can be used as a fuel subsequently. This process, called photocatalytic CO_2 reduction, uses the energy from sunlight to excite electrons

in the nanomaterial, which can then be used to reduce CO_2 into hydrocarbons. This process can potentially provide a solution to the problem of excess CO_2 in the atmosphere, while simultaneously producing renewable fuels.

In both the processes, the efficiency of the photocatalytic nanomaterial plays a crucial role in determining the overall energy storage performance. Therefore, researchers have been exploring ways to optimize the properties of photocatalytic nanomaterials to achieve higher efficiencies in energy storage applications. Some strategies include optimizing the band gap of nanomaterials, increasing their surface area and enhancing the charge separation efficiency. Photocatalytic nanomaterials have shown promising potential in energy storage applications such as hydrogen production and CO_2 reduction. With further research and development, photocatalytic nanomaterials can potentially provide a sustainable and clean solution to the world's energy needs.

8. Applications in Sensor Technology

Photocatalytic nanomaterials have shown great potential in the development of photodetectors and sensors due to their excellent optoelectronic properties and high surface area. The photocatalytic properties of these materials allow them to absorb light energy and generate charge carriers; this can be utilized in the detection of various analytes or to measure the intensity of light (Bai and Zhou 2014, Chaudhary et al. 2018, Rohilla et al. 2021, Zeng et al. 2014). Here are some of the ways in which photocatalytic nanomaterials can be used in photodetectors and sensors:

1. Gas sensors: Photocatalytic nanomaterials, by detecting changes in electrical conductivity, surface potential or surface resistance in the presence of specific gases, can be used as gas sensors. For example, metal oxide photocatalysts such as TiO_2, ZnO and SnO_2 have been used for the detection of various gases such as hydrogen, methane and nitrogen oxides.

2. Biosensors: Photocatalytic nanomaterials have also been used in biosensors which can detect biological molecules such as enzymes, proteins and DNA. Such biosensors utilize the unique properties of nanomaterials, such as their high surface area, to immobilize the biological molecules and generate a signal upon interaction with the target analyte.

3. Photodetectors: Photocatalytic nanomaterials have also been used in the development of photodetectors for detecting the intensity of light in various applications such as imaging, communication and sensing. For example, graphene-based photocatalysts have been used

to fabricate highly sensitive and efficient photodetectors due to their unique electronic and optoelectronic properties.

4. Environmental monitoring: Photocatalytic nanomaterials have also been used for environmental monitoring by detecting pollutants in air and water. For instance, metal oxide photocatalysts such as TiO_2 and ZnO have been used to detect the presence of pollutants such as volatile organic compounds (VOCs) and heavy metals in water and air.

5 Photovoltaic devices: Photocatalytic nanomaterials can be used in photovoltaic devices such as solar cells. By absorbing photons from sunlight, photocatalytic nanomaterials generate electron-hole pairs which can be subsequently separated and harnessed as electrical energy. This application has gained significant attention due to the potential of these materials for applications in renewable energy technologies.

6. Photoelectrochemical sensors: Photoelectrochemical sensors are sensors that use the photoelectric effect of photocatalytic nanomaterials to detect analytes in a solution. When exposed to light, the photocatalytic nanomaterial generates an electrical signal that changes depending on the presence and concentration of the analyte in the solution. This application has been utilized in various fields such as environmental monitoring, biomedical research and food safety.

7. Optical communication: Photocatalytic nanomaterials can also be used for optical communication, specifically for fiber-optic networks. By virtue of their light absorption and emission properties, photocatalytic nanomaterials can be used to convert light signals into electrical signals, allowing for faster and more efficient communication. With further research and development, these materials can potentially revolutionize the field of sensing and monitoring, providing solutions for various applications such as energy, communication, and environmental monitoring.

9. Applications in Environmental Technologies

Photocatalytic nanomaterials have significant potential for various applications in environmental technologies due to their unique properties and ability to catalyze chemical reactions upon exposure to light. These materials possess a high surface area to volume ratio, which increases their reactivity and provides a large number of active sites for chemical reactions to occur. When photocatalytic nanomaterials are exposed to light, they generate electron-hole pairs which can react with molecules in the surrounding environment to initiate various chemical reactions (Hu et al. 2021, Saber et al. 2020). This process is known as photocatalysis and is the basis of many environmental applications of photocatalytic nanomaterials. One of the key advantages of photocatalytic nanomaterials is their ability to degrade organic pollutants and harmful compounds, such as volatile organic compounds (VOCs), nitrogen oxides (NOx) and other toxic chemicals, into less harmful substances. This makes them particularly useful for air and water purification applications, where they can be incorporated into filters or coatings to remove pollutants and create cleaner, safer environments. In addition to their pollution reduction capabilities, photocatalytic nanomaterials also have potential for waste treatment, carbon dioxide reduction, soil remediation, disinfection and environmental monitoring applications (Ahmed and Haider 2018, Kandy and Fuels 2020, Mamba and Mishra 2016, Nahar et al. 2017, Perović et al. 2020, Wang et al. 2021a, Zhao et al. 2021). For example, they can be used to break down organic waste into simpler, less harmful compounds, convert carbon dioxide into useful chemicals, and detect pollutants in air, water and soil. Overall, the potential applications of photocatalytic nanomaterials in environmental technologies are numerous and varied. With further research and development, these materials hold great promise for addressing some of the most pressing environmental challenges we face today, such as air and water pollution, waste management and climate change. Some of the prominent and well researched application fields of photocatalytic nanomaterials in environmental technologies are covered below.

1. Air purification: Photocatalytic nanomaterials can be used for air purification by breaking down harmful pollutants such as volatile organic compounds (VOCs) and nitrogen oxides (NOx) through photocatalytic reactions. These materials can be incorporated into air purifiers, filters, or coatings on surfaces to reduce air pollution levels.

2. Water purification: Photocatalytic nanomaterials can also be used for water purification by degrading organic pollutants and killing bacteria and viruses through photocatalytic reactions. These materials can be incorporated into water treatment systems, filters, or coatings on surfaces to provide clean drinking water.

3. Waste treatment: Photocatalytic nanomaterials can also be used in waste treatment, specifically for the degradation of organic waste. The photocatalytic reaction can break down organic matter into simpler, less harmful compounds that can be more easily processed or recycled.

4. Self-cleaning surfaces: Photocatalytic nanomaterials can also be used to create self-cleaning surfaces, such as walls, floors and windows.

When a surface is coated with a photocatalytic nanomaterial, the surface can break down dirt and organic matter, allowing for easy cleaning and maintenance.

5. Soil remediation: Photocatalytic nanomaterials can be used for soil remediation by breaking down organic pollutants and eliminating pathogens. These materials can be used on contaminated soils and activated by sunlight to degrade the pollutants into less harmful substances.

6. Anti-fouling coatings: Photocatalytic nanomaterials can also be utilized as anti-fouling coatings on surfaces such as ship hulls, pipes and tanks. The photocatalytic reaction can break down organic matter and prevent the formation of biofilms and other organic accumulations, thus reducing maintenance and cleaning requirements.

7. Carbon dioxide reduction: Photocatalytic nanomaterials can be utilized for carbon dioxide reduction by converting carbon dioxide into useful chemicals such as methanol, methane or formic acid. These materials can be activated by sunlight to generate electron-hole pairs which can then drive the carbon dioxide reduction reaction.

8. Photocatalytic disinfection: Photocatalytic nanomaterials can be used for disinfection purposes by breaking down bacteria, viruses and other pathogens in air or water. These materials can be incorporated into air purifiers, water treatment systems or other disinfection equipment to provide clean and healthy environments.

9. Environmental monitoring: Photocatalytic nanomaterials can be used for environmental monitoring by detecting and measuring pollutants in air, water or soil. For example, nano sensors can be fabricated by incorporating photocatalytic nanomaterials into a sensing element, subsequently generating an electrical signal proportional to the concentration of the target pollutant.

Overall, photocatalytic nanomaterials have the potential to revolutionize environmental technologies by providing efficient and cost-effective solutions for air and water purification, waste treatment and self-cleaning surfaces. With further research and development, these materials can help address some of the world's most pressing environmental issues.

10. Nanomaterial-aided Photocatalysis: A Green Chemistry

Green chemistry refers to the design of chemical processes and products that reduce or eliminate the use and generation of hazardous substances. The use of photocatalytic nanomaterials in green chemistry has gained significant attention due to their ability to promote clean and sustainable chemical transformations by selective reactions, which means that they can catalyze specific chemical transformations without generating unwanted by-products or waste (Liu et al. 2014, Nunes et al. 2021, Yang et al. 2020, Zaleska-Medynska 2018). This is due to the fact that photocatalytic nanomaterials can be designed with specific shapes, sizes and compositions, which allows them to selectively interact with certain reactants and promote specific chemical reactions (Liu et al. 2014, Yang et al. 2020). Another important advantage of photocatalytic nanomaterials in green chemistry is their recyclability (Nunes et al. 2021). Because photocatalytic nanomaterials are not consumed during the catalytic reaction, they can be easily recovered and reused, thus reducing waste and the overall environmental impact of chemical reactions. Applications of photocatalytic nanomaterials in green chemistry also include the synthesis of fine chemicals, pharmaceuticals and natural products, as well as the degradation of pollutants and waste treatment. A carbon-free industrial application of photocatalytic nanomaterials is elucidated in Figure 3, exhibiting the production of carbon-free olefins using CO_2 captured from the environment. Moreover, photocatalytic reactions can be used for the synthesis of organic compounds such as aldehydes and ketones, which are commonly used as intermediates in the production of pharmaceuticals and other chemicals.

In addition to their applications in chemical synthesis, photocatalytic nanomaterials can also be used for environmental remediation, such as the degradation of pollutants in water and air (Bui et al. 2021, Gómez-Pastora et al. 2017, Mamba and Mishra 2016, Shen et al. 2015). This makes them an important tool for reducing the negative environmental impacts of industrial processes and improving the sustainability of chemical production. Here are some key examples of the green chemistry principles of photocatalysis:

1. Reducing the use of toxic chemicals: Photocatalytic reactions can often take place at room temperature and without a need of hazardous solvents or reagents, thus reducing the use of toxic chemicals and their environmental impact.

2. Energy efficiency: Photocatalysis is typically driven by renewable solar energy, hence reducing the need for energy from non-renewable sources and promoting energy efficiency.

3. Selective reactions: Photocatalytic nanomaterials can be designed to selectively catalyze specific chemical reactions, hence reducing the formation of unwanted by-products and waste.

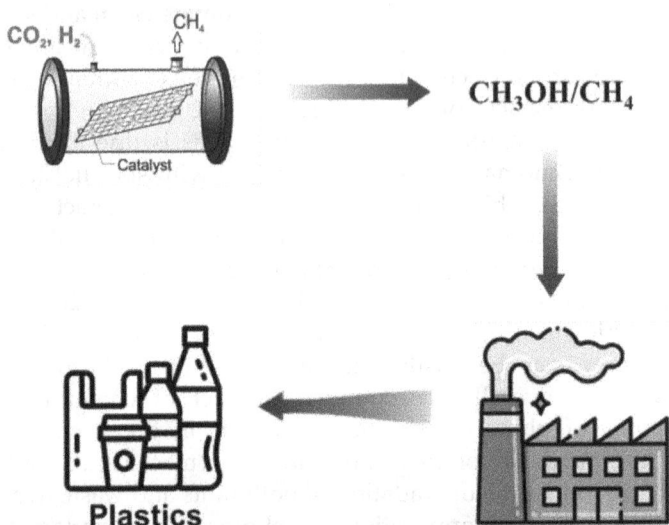

Figure 3: Green photocatalysis for the production of olefins, leading to the manufacture of plastics, by capturing atmospheric CO_2.

4. Recyclability: Photocatalytic nanomaterials can be recycled and reused, which reduces the amount of waste generated during chemical reactions.

5. Safety: The use of photocatalytic nanomaterials in green chemistry is generally considered safe due to their low toxicity and good biocompatibility.

Ultimately, the use of photocatalytic nanomaterials in green chemistry represents a significant step forward in creating more sustainable and environment friendly chemical processes and products. With further research and development, it is certain that photocatalytic nanomaterials will continue to play a key role in advancing the field of green chemistry and promoting more sustainable chemical practices.

11. Thermodynamic/kinetic Stability and Market Penetrability of Photocatalytic Nanomaterials

Thermodynamic stability of photocatalytic nanomaterials refers to their ability to maintain their chemical and physical properties over time and under various environmental conditions. The stability of these materials is important for their reliable and effective use in various applications, including photocatalysis (Andrievski 2014). Thermodynamic stability is typically evaluated through various analytical techniques, including X-ray diffraction, transmission electron microscopy and spectroscopy methods

(Lin et al. 2014). These techniques can provide information about the crystal structure, morphology and chemical composition of photocatalytic nanomaterials, which can in turn help determine their stability (Zhou et al. 2022). Factors that can affect the thermodynamic stability of photocatalytic nanomaterials include their composition, size, shape and surface properties (Xu et al. 2018). For example, materials with high surface-area-to-volume ratio, such as nanoparticles, are more susceptible to oxidation and degradation compared to bulk materials. Additionally, the presence of impurities or defects can also affect the stability of photocatalytic nanomaterials. Improving the thermodynamic stability of photocatalytic nanomaterials is an active area of research as it can lead to more efficient and durable materials for use in various applications (Marks and Peng 2016). Strategies for improving stability include modifying the surface chemistry of these materials, optimizing their size and shape, and using protective coatings or encapsulation methods (Marks and Peng 2016). Overall, understanding and improving the thermodynamic stability of photocatalytic nanomaterials is crucial for their effective use in various applications, including environmental remediation, energy production and storage, and biomedical applications.

Kinetic stability of photocatalytic nanomaterials is their ability to resist changes in their chemical and physical properties over time and under various environmental conditions. Unlike thermodynamic stability, which refers to the equilibrium properties of a material, kinetic stability refers to the resistance of a material to chemical reactions or degradation over time (Xu et al. 2018). Factors that can affect the kinetic stability of photocatalytic nanomaterials include their composition, size, shape, surface area and exposure to light and reactive species (Liu et al. 2022). In particular, the presence of reactive species, such as oxygen, can lead to the degradation of photocatalytic nanomaterials over time. To improve the kinetic stability of photocatalytic nanomaterials, various strategies can be employed, including optimizing their size and shape, modifying their surface chemistry, and using protective coatings or encapsulation methods (Vandenabeele et al. 2020). For example, modifying the surface chemistry of photocatalytic nanomaterials can help reduce their susceptibility to reactive species, while using protective coatings or encapsulation methods can help prevent degradation and enhance their stability (Görke and Garnweitner 2021, Wen et al. 2015). The kinetic stability of photocatalytic nanomaterials is an important consideration for their use in various applications, including environmental remediation, energy production and storage, and biomedical applications. By improving their kinetic stability, photocatalytic nanomaterials can provide more durable and effective solutions for these applications.

Despite their promising potential for various industrial chemistry applications, the commercial feasibility of these advanced materials is still under evaluation, as many factors must be considered before deploying these materials on an industrial scale. One major consideration for the commercial feasibility of photocatalytic nanomaterials is their production cost (McINTYRE 2012). Currently, the production of these materials can be expensive, which may limit their widespread adoption in industry. Therefore, research efforts are focused on developing more cost-effective and scalable methods for the synthesis of photocatalytic nanomaterials. Another important factor is the stability and durability of photocatalytic nanomaterials under industrial operating conditions (Moon et al. 2016). The stability of these materials is critical for their effective and reliable use in various industrial applications (Huang et al. 2020). Therefore, research efforts are focused on improving the stability of photocatalytic nanomaterials by developing protective coatings, encapsulation methods and surface modifications (Akhoondi et al. 2021, Görke and Garnweitner 2021, Valizadeh et al. 2018, Vengatesan et al. 2015). The efficiency of photocatalytic nanomaterials is also a crucial factor for their commercial feasibility in industrial chemistry. The efficiency of these materials is influenced by their size, shape, composition and surface area, among other factors (Bradley et al. 2011, Hansen and Baun 2012). Therefore, research efforts are focused on optimizing the properties of photocatalytic nanomaterials to enhance their efficiency and performance. However, there are several industrial chemistry applications where photocatalytic nanomaterials have shown promise, including air and water purification, wastewater treatment and energy production (Gautam et al. 2020, Gómez-Pastora et al. 2017, Nahar et al. 2017, Nunes et al. 2021, Perović et al. 2020, Ren et al. 2017, Santos et al. 2015, Yan et al. 2018, Zhao et al. 2021). Research and development efforts in this field are ongoing, and with continued progress in the synthesis and optimization of photocatalytic nanomaterials, their commercial feasibility in industrial chemistry is likely to increase in the future. The market penetration potential of photocatalytic nanomaterials in industrial chemistry is significant, as these materials have the potential to provide sustainable and cost-effective solutions for various industrial applications. In recent years, there has been an increasing demand for sustainable and environment friendly technologies in the industrial sector. Photocatalytic nanomaterials have emerged as a promising solution to address this demand, as they have shown great potential in various applications, including air and water purification, wastewater treatment and energy production. Table 1 shows the photocatalytic nanomaterials successfully and widely adopted in the market in recent years.

Table 1: Photocatalytic nanomaterials successfully adopted by the market in recent years.

Order	Industrial name	Synthesis method	Composition	Details/application	Ref.
1.	Fe-MIL-100	microwave-assisted	Metal(III)-carboxylate-based MOFs	nanoscale particles; drug delivery and imaging	(Horcajada et al. 2010)
2.	Fe-MIL-101	microwave-assisted	Metal(III)-carboxylate-based MOFs	nanoscale particles	(Taylor-Pashow et al. 2009)
3.	Cr-MIL-101	microwave-assisted	Metal(III)-carboxylate-based MOFs	nanoscale particles; benzene sorption	(Jhung et al. 2007)
4.	IRMOF-n	microwave-assisted	Metal(II)-carboxylate-based MOFs	IRMOF-1 membrane	(Yoo et al. 2009)
5.	IRMOF-n	microwave-assisted	Metal(II)-carboxylate-based MOFs	IRMOF-2, -3	(Ni and Masel, 2006)
6.	ZIF-8	microwave-assisted	Azolate-based MOFS	ZIF-8 powder	(Park et al. 2009)
7.	[Zn2(NDC)2(DPNI)]	microwave-assisted	Mixed-linker MOFs	selective gas sorption	(Bae et al. 2008)
8.	Zn- and Cu-carboxylates	Electrochemical		systematic investigation of Zn, Cu, Mg and Co as anode materials; 1,3,5-H3BTC, 1,2,3-H3BTC, H2BDC and H2BDC-(OH)2 as linkers	(Mueller et al. 2007)
9.	HKUST-1	Electrochemical		synthesis; gas purification, storage and separation	(Mueller et al. 2006)
10.	HKUST-1	Electrochemical		synthesis; isobutene-isobutane separation	(Hartmann et al. 2008)
11.	HKUST-1	Electrochemical		patterned film growth	(Ameloot et al. 2009)
12.	HKUST-1	Electrochemical		patterned film growth by galvanic displacement	(Ameloot et al. 2010)
13.	[Cu(INA)2]	mechanochemical	Carboxylate-based MOFs	solvent-free synthesis; known phase	(Pichon et al. 2006)

Table 1 contd. ...

...Table 1 contd.

Order	Industrial name	Synthesis method	Composition	Details/application	Ref.
14.	HKUST-1	mechanochemical	Carboxylate-based MOFs	H2 storage	(Yang et al. 2011)
15.	[Zn(EIm)2]	mechanochemical	Imidazolate-based MOFs	dense and porous ZIFs; ILAG; ZnO as starting material; metal salts as structure-directing agents	(Beldon et al. 2010)
16.	[Co3Cl6(L')2]	mechanochemical	Imidazolate-based MOFs	tripodal imidazole frameworks; manual grinding	(Willans et al. 2011)
17.	[Zn2(FMA)2(BPY)]	mechanochemical	Mixed-linker, pillared-layered MOFs	varying amount of AcOH and H2O in the pores; solvent-free grinding	(Fujii et al. 2010)
18.	MOF-5	Sonochemical	Ditopic linker	statistical design used for synthesis optimization; consecutive application of US and MW irradiation	(Sabouni et al. 2010)
19.	IRMOF-9 and -10	Sonochemical	Ditopic linker	control of catenation and its effect on CO2 adsorption; role of power level in US synthesis	(Kim et al. 2011)
20.	Fe-MIL-88A	Sonochemical	Ditopic linker	nanoparticles; flexible porous MOF; CE, MW and US synthesis	(Chalati et al. 2011)
21.	HKUST-1	Sonochemical	Tritopic Linker	time-dependent formation in DMF/ethanol/water mixture	(Li et al. 2009)
22.	MOF-177	Sonochemical	Tritopic Linker	comparison among CE, MW and US syntheses; short reaction time; high yield; small particles; improved sorption properties	(Jung et al. 2010)
23.	PCN-6 and -60	Sonochemical	Tritopic Linker	control of catenation and its effect on CO2 adsorption; role of power level in US synthesis	(Kim et al. 2011)

Moreover, the market for photocatalytic nanomaterials is expected to grow in the coming years, as more industries seek sustainable and cost-effective solutions for their operations. The demand for these materials is likely to increase in industries such as chemical processing, food and beverage production, pharmaceuticals, textiles and several other industries in addition to industrial applications discussed above. However, the market penetration potential of photocatalytic nanomaterials in industrial chemistry is also influenced by factors such as production costs, regulatory requirements and competing technologies. As mentioned earlier, the production costs of these materials can be high, which may limit their adoption in some industries. Additionally, regulatory requirements must be met to ensure the safe and effective use of photocatalytic nanomaterials in industrial settings. Overall, the market penetration potential of photocatalytic nanomaterials in industrial chemistry is promising, and with continued research and development efforts, their adoption in various industrial applications is likely to increase in the future.

12. Conclusion, Outlook and Future Challenges

Photocatalytic nanomaterials have emerged as a promising solution for many challenges faced by various industries and fields, including energy, environment, sensor technology and healthcare. Research efforts in recent years have focused on improving the efficiency, selectivity and stability of these materials, as well as developing new fabrication methods and surface modification techniques to enhance their properties. The research and field applications of photocatalytic nanomaterials are expected to continue to grow in the coming years. Some of the key areas of focus include: (1) development of new materials: researchers are constantly working to develop new and improved photocatalytic nanomaterials with better properties, increased stability, and more cost-effective production methods; this will lead to a wider range of applications in various fields; (2) environmental applications: photocatalytic nanomaterials are being increasingly used in environmental applications, such as water treatment, air purification, and waste management; as environmental concerns continue to grow, the demand for these materials is expected to increase; (3) energy applications: photocatalytic nanomaterials have great potential in energy-related applications such as solar energy conversion and energy storage; as the world continues to shift towards renewable energy sources, these materials will play an important role in enabling the transition; (4) sensor technology: photocatalytic nanomaterials are also being used in the development of new sensor technologies, which have a wide range of applications in areas such as healthcare, food safety and environmental monitoring; (5) commercialization: with increasing demand for photocatalytic nanomaterials in various fields, more efforts are being made to commercialize these materials; this includes developing new manufacturing processes, improving scalability and reducing costs. Overall, the future looks bright for photocatalytic nanomaterials as research continues to drive new advancements and applications in various fields. With a continued focus on improving the properties and production methods of these materials, and increasing public awareness and acceptance, the potential for their widespread use and impact will continue to grow. While the potential applications of photocatalytic nanomaterials are vast, there are still several challenges that need to be addressed to ensure their successful adoption in various fields. Some of the key challenges are:

1. Cost-effectiveness: The production cost of photocatalytic nanomaterials is still high, which can limit their adoption in certain industrial applications. Research efforts are needed to develop more cost-effective production methods.

2. Stability: The long-term stability of photocatalytic nanomaterials is a concern, for their properties can deteriorate over time, affecting their efficiency and effectiveness in various applications. More research is needed to understand and improve the stability of these materials.

3. Scalability: Scaling up the production of photocatalytic nanomaterials from the laboratory scale to industrial scale is a challenge. Optimizing the production processes for industrial-scale manufacturing while maintaining quality is essential for the widespread adoption of these materials. However, with the increasing demand for clean and sustainable technologies, there is a growing interest from both public and private sectors to develop and commercialize photocatalytic nanomaterials.

4. Standardization: Standardization of photocatalytic nanomaterials is necessary to ensure their safety, efficacy and compatibility with different applications. Developing standardized testing protocols and regulatory guidelines will help accelerate their adoption in various fields. Moreover, there is a need for standardized methods to evaluate the safety and toxicity of photocatalytic nanomaterials, particularly in healthcare applications.

5. Integration: Integration of photocatalytic nanomaterials into existing industrial processes and systems can be challenging. Research efforts are required to develop methods for integrating these materials into existing systems while minimizing the disruption of the existing processes.

6. Public acceptance: The safety and the environmental impacts of photocatalytic nanomaterials are still not fully understood, and public

acceptance of these materials can be a barrier to their adoption in various fields. More research is needed to understand their safety and environmental impacts and to communicate this information to the public. However, researchers are exploring sustainable synthesis methods and developing strategies for recycling and reusing these materials.

While photocatalytic nanomaterials have great potential for various applications in industrial chemistry, environmental technologies, sensor technology and other fields, addressing the abovementioned challenges is crucial for their successful adoption and widespread use in these fields. However, despite these challenges, the outlook towards photocatalytic nanomaterials is positive, with numerous research groups, industries and governments investing in this field. Continued research and collaborations among stakeholders will be essential to further advance the field and unlock the full potential of these materials in various applications.

References

Abdullah, F., Bakar, N.A. and Bakar, M.A.J.J.o.h.m. 2022. Current advancements on the fabrication, modification, and industrial application of zinc oxide as photocatalyst in the removal of organic and inorganic contaminants in aquatic systems. Journal of Hazardous Materials, 424: 127416.

Ahmed, S.N. and Haider, W.J.N. 2018. Heterogeneous photocatalysis and its potential applications in water and wastewater treatment: A review. Nanotechnology, 29(34): 342001.

Akhoondi, A., Feleni, U., Bethi, B., Idris, A.O., Hojjati-Najafabadi, A.J.S. and Sintering. 2021. Advances in metal-based vanadate compound photocatalysts: Synthesis, properties and applications. Synthesis and Sintering, 1(3): 151–168.

Ali, M., Anjum, A.S., Bibi, A., Wageh, S., Sun, K.C. and Jeong, S.H.J.C. 2022. Gradient heating-induced bi-phase synthesis of carbon quantum dots (CQDs) on graphene-coated carbon cloth for efficient photoelectrocatalysis. Carbon, 196: 649–662.

Ameloot, R., Pandey, L., Van der Auweraer, M., Alaerts, L., Sels, B.F. and De Vos, D.E. 2010. Patterned film growth of metal–organic frameworks based on galvanic displacement. Chemical Communications, 46(21): 3735–3737.

Ameloot, R., Stappers, L., Fransaer, J., Alaerts, L., Sels, B.F. and De Vos, D.E. 2009. Patterned growth of metal-organic framework coatings by electrochemical synthesis. Chemistry of Materials, 21(13): 2580–2582.

Andrievski, R.J.J.o.m.s. 2014. Review of thermal stability of nanomaterials. Journal of Material Science, 49: 1449–1460.

Ansari, A.A., Sillanpää, M.J.R. and Reviews, S.E. 2021. Advancement in upconversion nanoparticles based NIR-driven photocatalysts. Renewable and Sustainable Energy Reviews, 151: 111631.

Bae, Y.-S., Mulfort, K.L., Frost, H., Ryan, P., Punnathanam, S., Broadbelt, L.J. et al. 2008. Separation of CO2 from CH4 using mixed-ligand metal–organic frameworks. Langmuir, 24(16): 8592–8598.

Bai, J. and Zhou, B.J.C.r. 2014. Titanium dioxide nanomaterials for sensor applications. Chemical Reviews, 114(19): 10131–10176.

Beldon, P.J., Fábián, L., Stein, R.S., Thirumurugan, A., Cheetham, A.K. and Friščić, T. 2010. Rapid room-temperature synthesis of zeolitic imidazolate frameworks by using mechanochemistry. Angewandte Chemie, 122(50): 9834–9837.

Belousov, A.S. and Suleimanov, E.V.J.G.C. 2021. Application of metal–organic frameworks as an alternative to metal oxide-based photocatalysts for the production of industrially important organic chemicals. Green Chemistry, 23(17): 6172–6204.

Bradley, E.L., Castle, L. and Chaudhry, Q.J.T.i.f.s., technology. 2011. Applications of nanomaterials in food packaging with a consideration of opportunities for developing countries. Trends in Food Science & Technology, 22(11): 604–610.

Bui, V.K.H., Nguyen, T.N., Van Tran, V., Hur, J., Kim, I.T., Park, D. et al. Innovation. 2021. Photocatalytic materials for indoor air purification systems: An updated mini-review. Environmental Technology & Innovation, 22: 101471.

Chalati, T., Horcajada, P., Gref, R., Couvreur, P. and Serre, C. 2011. Optimisation of the synthesis of MOF nanoparticles made of flexible porous iron fumarate MIL-88A. Journal of Materials Chemistry, 21(7): 2220–2227.

Chaudhary, S., Umar, A., Bhasin, K. and Baskoutas, S.J.M. 2018. Chemical sensing applications of ZnO nanomaterials. Materials, 11(2): 287.

Chen, H., Liu, K., Hu, L., Al-Ghamdi, A.A. and Fang, X.J.M.T. 2015. New concept ultraviolet photodetectors. Materials Today, 18(9): 493–502.

Chen, Y., Lai, Z., Zhang, X., Fan, Z., He, Q., Tan, C. et al. 2020. Phase engineering of nanomaterials. Nature Reviews Chemistry, 4(5): 243–256.

Cipolatti, E.P., Valerio, A., Henriques, R.O., Moritz, D.E., Ninow, J.L., Freire, D.M. et al. 2016. Nanomaterials for biocatalyst immobilization–state of the art and future trends. RSC Advances, 6(106): 104675–104692.

Donati, S.J.M.S. and Technology. 2001. Photodetectors: Devices, circuits, and applications. Measurement Science and Technology, 12(5): 653–663.

Fujii, K., Garay, A.L., Hill, J., Sbircea, E., Pan, Z., Xu, M. et al.2010. Direct structure elucidation by powder X-ray diffraction of a metal–organic framework material prepared by solvent-free grinding. Chemical Communications, 46(40): 7572–7574.

Gautam, S., Agrawal, H., Thakur, M., Akbari, A., Sharda, H., Kaur, R. et al. 2020. Metal oxides and metal organic frameworks for the photocatalytic degradation: A review. Journal of Environmental Chemical Engineering, 8(3): 103726.

Gómez-Pastora, J., Dominguez, S., Bringas, E., Rivero, M.J., Ortiz, I. and Dionysiou, D.D.J.C.E.J. 2017. Review and perspectives on the use of magnetic nanophotocatalysts (MNPCs) in water treatment. Chemical Engineering Journal, 310: 407–427.

Görke, M. and Garnweitner, G.J.C. 2021. Crystal engineering of nanomaterials: Current insights and prospects. CrystEngComm, 23(45): 7916–7927.

Hansen, S.F. and Baun, A.J.D.-R. 2012. European regulation affecting nanomaterials-review of limitations and future recommendations. Dose Response, 10(3): 10–029.

Hao, N., Li, L. and Tang, F.J.I.M.R. 2017. Roles of particle size, shape and surface chemistry of mesoporous silica nanomaterials on biological systems. International Materials Reviews, 62(2): 57–77.

Hartmann, M., Kunz, S., Himsl, D., Tangermann, O., Ernst, S. and Wagener, A. 2008. Adsorptive separation of isobutene and isobutane on Cu3 (BTC) 2. Langmuir, 24(16): 8634–8642.

Horcajada, P., Chalati, T., Serre, C., Gillet, B., Sebrie, C., Baati, T. et al. 2010. Porous metal–organic-framework nanoscale carriers as a potential platform for drug delivery and imaging. Nature Materials, 9(2): 172.

Hou, D., Zhou, W., Liu, X., Zhou, K., Li, G., Chen, S.J.E. et al. 2014. Two-dimensional photocatalysts: Properties, synthesis, and applications. Energy and Environment Focus, 3(4): 330–338.

Hu, G., Yang, J., Duan, X., Farnood, R., Yang, C., Yang, J. et al. 2021. Recent developments and challenges in zeolite-based composite photocatalysts for environmental applications. Chemical Engineering Journal, 417: 129209.

Huang, H., Pradhan, B., Hofkens, J., Roeffaers, M.B. and Steele, J.A.J.A.E.L. 2020. Solar-driven metal halide perovskite photocatalysis: design, stability, and performance. ACS Energy Letters, 5(4): 1107–1123.

Hussain, C.M. 2020. Handbook of Functionalized Nanomaterials for Industrial Applications. Elsevier.

Jhung, S.H., Lee, J.H., Yoon, J.W., Serre, C., Férey, G. and Chang, J.S. 2007. Microwave synthesis of chromium terephthalate MIL-101 and its benzene sorption ability. Advanced Materials, 19(1): 121–124.

Jung, D.-W., Yang, D.-A., Kim, J., Kim, J. and Ahn, W.-S. 2010. Facile synthesis of MOF-177 by a sonochemical method using 1-methyl-2-pyrrolidinone as a solvent. Dalton Transactions, 39(11): 2883–2887.

Kandy, M.M.J.S.E., Fuels. 2020. Carbon-based photocatalysts for enhanced photocatalytic reduction of CO_2 to solar fuels. Sustainable Energy & Fuels, 4(2): 469–484.

Karthik, V., Selvakumar, P., Senthil Kumar, P., Vo, D.-V.N., Gokulakrishnan, M., Keerthana, P. et al. 2021. Graphene-based materials for environmental applications: A review. Environmental Chemistry Letters, 19(5): 3631–3644.

Karthikeyan, C., Arunachalam, P., Ramachandran, K., Al-Mayouf, A.M., Karuppuchamy, S.J.J.o.A., Compounds. 2020. Recent advances in semiconductor metal oxides with enhanced methods for solar photocatalytic applications. Journal of Alloys and Compounds, 828: 154281.

Kim, J., Yang, S.-T., Choi, S.B., Sim, J., Kim, J. and Ahn, W.-S. 2011. Control of catenation in CuTATB-n metal–organic frameworks by sonochemical synthesis and its effect on CO_2 adsorption. Journal of Materials Chemistry, 21(9): 3070–3076.

Kolahalam, L.A., Viswanath, I.K., Diwakar, B.S., Govindh, B., Reddy, V. and Murthy, Y.J.M.T.P. 2019. Review on nanomaterials: Synthesis and applications. Materisls Today: Proceedings, 18: 2182–2190.

Lettieri, S., Pavone, M., Fioravanti, A., Santamaria Amato, L. and Maddalena, P.J.M. 2021. Charge carrier processes and optical properties in TiO_2 and TiO_2-based heterojunction photocatalysts: A review. Materials, 14(7): 1645.

Li, Z.-Q., Qiu, L.-G., Xu, T., Wu, Y., Wang, W., Wu, Z.-Y. et al. 2009. Ultrasonic synthesis of the microporous metal–organic framework Cu3 (BTC) 2 at ambient temperature and pressure: an efficient and environmentally friendly method. Materials Letters, 63(1): 78–80.

Lin, P.-C., Lin, S., Wang, P.C. and Sridhar, R.J.B.a. 2014. Techniques for physicochemical characterization of nanomaterials. Biotechnology Advances, 32(4): 711–726.

Liu, L., Wang, S., Jiang, G., Zhang, B., Yang, J., Wang, J. et al. 2022. Solvothermal synthesis of zirconia nanomaterials: Latest developments and future.

Liu, S., Zhang, N., Xu, Y.J.J.P. and Characterization, P.S. 2014. Core–shell structured nanocomposites for photocatalytic selective organic transformations. Particle & Particle System Characterization, 31(5): 540–556.

Long, M., Wang, P., Fang, H. and Hu, W.J.A.F.M. 2019. Progress, challenges, and opportunities for 2D material based photodetectors. Advanced Functional Materials, 29(19): 1803807.

Lopez, Y.C., Viltres, H., Gupta, N.K., Acevedo-Pena, P., Leyva, C., Ghaffari, Y. et al. 2021. Transition metal-based metal–organic frameworks for environmental applications: A review. Environmental Chemistry Letters, 19: 1295–1334.

Mamba, G. and Mishra, A.J.A.C.B.E. 2016. Graphitic carbon nitride (g-C3N4) nanocomposites: a new and exciting generation of visible light driven photocatalysts for environmental pollution remediation. Applied Catalysis B: Environmental, 198: 347–377.

Marks, L. and Peng, L.J.J.o.P.C.M. 2016. Nanoparticle shape, thermodynamics and kinetics. Journal of Physics: Condensed Matter, 28(5): 053001.

McINTYRE, R.A.J.S.p. 2012. Common nano-materials and their use in real world applications. Science Progress, 95(1): 1–22.

Mishra, R., Bera, S., Chatterjee, R., Banerjee, S., Bhattacharya, S., Biswas, A. et al. 2022. A review on Z/S-scheme heterojunction for photocatalytic applications based on metal halide perovskite materials. Applied Surface Science Advances, 9: 100241.

Moon, R.J., Schueneman, G.T. and Simonsen, J.J.J. 2016. Overview of cellulose nanomaterials, their capabilities and applications. Jom, 68: 2383–2394.

Mueller, U., Puetter, H., Hesse, M. and Wessel, H. 2007. WO 2005/049892, 2005. BASF Aktiengesellschaft.

Mueller, U., Schubert, M., Teich, F., Puetter, H., Schierle-Arndt, K. and Pastre, J. 2006. Metal–organic frameworks—prospective industrial applications. Journal of Materials Chemistry, 16(7): 626–636.

Nahar, S., Zain, M., Kadhum, A.A.H., Hasan, H.A. and Hasan, M.R.J.M. 2017. Advances in photocatalytic CO_2 reduction with water: A review. Materials, 10(6): 629.

Ni, Z. and Masel, R.I. 2006. Rapid production of metal– organic frameworks via microwave-assisted solvothermal synthesis. Journal of the American Chemical Society, 128(38): 12394–12395.

Nunes, D., Pimentel, A., Branquinho, R., Fortunato, E. and Martins, R.J.C. 2021. Metal oxide-based photocatalytic paper: A green alternative for environmental remediation. Catalysts, 11(4): 504.

Park, J.-H., Park, S.-H. and Jhung, S.-H. 2009. Microwave-syntheses of zeolitic imidazolate framework material, ZIF-8. Journal of the Korean Chemical Society, 53(5): 553–559.

Perović, K., dela Rosa, F.M., Kovačić, M., Kušić, H., Štangar, U.L., Fresno, F. et al. 2020. Recent achievements in development of TiO2-based composite photocatalytic materials for solar driven water purification and water splitting. Materials, 13(6): 1338.

Pichon, A., Lazuen-Garay, A. and James, S.L. 2006. Solvent-free synthesis of a microporous metal–organic framework. CrystEngComm, 8(3): 211–214.

Putzbach, W. and Ronkainen, N.J.J.S. 2013. Immobilization techniques in the fabrication of nanomaterial-based electrochemical biosensors: A review. Sensors, 13(4): 4811–4840.

Ren, H., Koshy, P., Chen, W.-F., Qi, S. and Sorrell, C.C.J.J.o.H.M. 2017. Photocatalytic materials and technologies for air purification. Journal of Hazardous Materials, 325: 340–366.

Riaz, R., Ali, M., Maiyalagan, T., Anjum, A.S., Lee, S., Ko, M.J. et al. 2019. Dye-sensitized solar cell (DSSC) coated with energy down shift layer of nitrogen-doped carbon quantum dots (N-CQDs) for enhanced current density and stability. Applied Surface Science, 483: 425–431.

Rohilla, D., Chaudhary, S. and Umar, A.J.E.S. 2021. An overview of advanced nanomaterials for sensor applications. Engineered Science, 16: 47–70.

Saber, N.B., Mezni, A., Alrooqi, A., Altalhi, T.J.J.o.M.R., Technology. 2020. A review of ternary nanostructures based noble metal/semiconductor for environmental and renewable energy applications. Journal of Materials Research and Technology, 9(6): 15233–15262.

Sabouni, R., Kazemian, H. and Rohani, S. 2010. A novel combined manufacturing technique for rapid production of IRMOF-1 using ultrasound and microwave energies. Chemical Engineering Journal, 165(3): 966–973.

Sakdaronnarong, C., Sangjan, A., Boonsith, S., Kim, D.C. and Shin, H.S.J.C. 2020. Recent developments in synthesis and photocatalytic applications of carbon dots. Catalysts, 10(3): 320.

Santos, C.S., Gabriel, B., Blanchy, M., Menes, O., García, D., Blanco, M. et al. 2015. Industrial applications of nanoparticles—A prospective overview. Materials Today: Proceedings, 2(1): 456–465.

Shen, Y., Fang, Q. and Chen, B.J.E.S., Technology. 2015. Environmental applications of three-dimensional graphene-based macrostructures: Adsorption, transformation, and detection. Environmental Science and Technology, 49(1): 67–84.

Shuai, W., Das, R.K., Naghdi, M., Brar, S.K., Verma, M.J.B., biochemistry. 2017. A review on the important aspects of lipase immobilization on nanomaterials. Biotechnology and Applied Biochemistry, 64(4): 496–508.

Taylor-Pashow, K.M., Della Rocca, J., Xie, Z., Tran, S. and Lin, W. 2009. Postsynthetic modifications of iron-carboxylate nanoscale metal–organic frameworks for imaging and drug delivery. Journal of the American Chemical Society, 131(40): 14261–14263.

Tran, D.P., Pham, M.-T., Bui, X.-T., Wang, Y.-F. and You, S.-J.J.S.E. 2022. CeO$_2$ as a photocatalytic material for CO$_2$ conversion: A review. Solar Energy, 240: 443–466.

Valizadeh, B., Nguyen, T.N. and Stylianou, K.C.J.P. 2018. Shape engineering of metal–organic frameworks. Polyhedron, 145: 1–15.

Vandenabeele, C.R., Lucas, S.J.M.S. and Reports, E.R. 2020. Technological challenges and progress in nanomaterials plasma surface modification–A review. Materials Science and Engineering, 139: 100521.

Vengatesan, M.R., Mittal, V.J.S.m.o.n. and fillers, n.f. 2015. Surface modification of nanomaterials for application in polymer nanocomposites: An overview. Surface Modification of Nanoparticles and Natural Fiber Fillers, 1–28.

Wang, D., Mueses, M.A., Márquez, J.A.C., Machuca-Martínez, F., Grčić, I., Moreira, R.P.M. et al. 2021a. Engineering and modeling perspectives on photocatalytic reactors for water treatment. Water Research, 202: 117421.

Wang, Y., Guo, H., Luo, X., Liu, X., Hu, Z., Han, L. et al. 2019. Nonsiliceous mesoporous materials: design and applications in energy conversion and storage. Small, 15(32): 1805277.

Wang, Y., Wang, L., Zhang, X., Liang, X., Feng, Y. and Feng, W.J.N.T. 2021b. Two-dimensional nanomaterials with engineered bandgap: Synthesis, properties, applications. Nano Today, 37: 101059.

Wen, J., Li, X., Liu, W., Fang, Y., Xie, J., Xu and Y.J.C.J.o.C. 2015. Photocatalysis fundamentals and surface modification of TiO$_2$ nanomaterials. Chinese Journal of Catalysis, 36(12): 2049–2070.

Willans, C.E., French, S., Anderson, K.M., Barbour, L.J., Gertenbach, J.-A., Lloyd, G.O. et al. 2011. Tripodal imidazole frameworks: Reversible vapour sorption both with and without significant structural changes. Dalton Transactions, 40(3): 573–582.

Wu, W., Jiang, C. and Roy, V.A.J.N. 2015. Recent progress in magnetic iron oxide–semiconductor composite nanomaterials as promising photocatalysts. Nanoscale, 7(1): 38–58.

Xu, L., Liang, H.-W., Yang, Y. and Yu, S.-H.J.C.r. 2018. Stability and reactivity: Positive and negative aspects for nanoparticle processing. Chemical Reviews, 118(7): 3209–3250.

Yadav, S. and Jaiswar, G.J.J.o.t.C.C.S. 2017. Review on undoped/doped TiO2 nanomaterial; synthesis and photocatalytic and antimicrobial activity. Journal of the Chinese Chemical Society, 64(1): 103–116.

Yamashita, H. and Li, H. 2016. Nanostructured Photocatalysts. Springer.

Yan, J., Verma, P., Kuwahara, Y., Mori, K. and Yamashita, H.J.S.M. 2018. Recent progress on black phosphorus-based materials for photocatalytic water splitting. Small Methods, 2(12): 1800212.

Yang, H., Orefuwa, S. and Goudy, A. 2011. Study of mechanochemical synthesis in the formation of the metal–organic framework Cu3 (BTC) 2 for hydrogen storage. Microporous and Mesoporous Materials, 143(1): 37–45.

Yang, X., Zhang, S., Li, P., Gao, S. and Cao, R.J.J.o.M.C.A. 2020. Visible-light-driven photocatalytic selective organic oxidation reactions. Journal of Materials Chemistry A, 8(40): 20897–20924.

Yoo, Y., Lai, Z. and Jeong, H.-K. 2009. Fabrication of MOF-5 membranes using microwave-induced rapid seeding and solvothermal secondary growth. Microporous and Mesoporous Materials, 123(1-3): 100–106.

Zaleska-Medynska, A. 2018. Metal Oxide-based Photocatalysis: Fundamentals and Prospects for Application. Elsevier.

Zeng, S., Baillargeat, D., Ho, H.-P. and Yong, K.-T.J.C.S.R. 2014. Nanomaterials enhanced surface plasmon resonance for biological and chemical sensing applications. Chemical Society Reviews, 43(10): 3426–3452.

Zhao, G.Q., Hu, J., Long, X., Zou, J., Yu, J.G. and Jiao, F.P.J.S. 2021. A critical review on black phosphorus-based photocatalytic CO_2 reduction application. Small, 17(49): 2102155.

Zhao, W., Adeel, M., Zhang, P., Zhou, P., Huang, L., Zhao, Y. et al. 2022. A critical review on surface-modified nano-catalyst application for the photocatalytic degradation of volatile organic compounds. Environmental Science: Nano, 9(1): 61–80.

Zheng, Z., Cox, M. and Li, B.J.J.o.M.S. 2018. Surface modification of hexagonal boron nitride nanomaterials: A review. Journal of Material Science, 53: 66–99.

Zhou, Q., Duan, J., Duan, Y. and Tang, Q.J.J.o.E.C. 2022. Review on engineering two-dimensional nanomaterials for promoting efficiency and stability of perovskite solar cells. Journal of Energy Chemistry, 68: 154–175.

CHAPTER 4

Application of Functionalized Nanomaterials for Wastewater Restoration

Shakeel Ahmad,[1,2] Jingchun Tang,[1,] Xiaomei Liu,[3]
Linan Liu,[1] Muhammad Azher Hassan[4] and Tariq Mehmood[5]*

1. Introduction

Of the total water on Earth, 97.5% is considered saline water and 2.5% is fresh water. Around 68.9% of the said fresh water is found as ice, snow and glaciers, 30.8% as groundwater, and only 0.3% is accessible for our use (Jain et al. 2021). The demand for freshwater has shown to increase by about 70%, 22% and 8% respectively for agriculture, industry and domestic consumption (Helmer and Hespanhol 1997). As an ultimate effluent of fresh water, wastewater comprises many hazardous and harmful substances and comes from many sources, i.e., domestic, industrial, hospital and agricultural discharges and storm runoff; they can be characterized by their physical appearance, chemical composition and microbial loads. Wastewater pollutants are the reason for multiple

[1] MOE Key Laboratory of Pollution Processes and Environmental Criteria/Tianjin Engineering Center of Environmental Diagnosis and Contamination Remediation, College of Environmental Science and Engineering, Nankai University, Tianjin 300350, China.
[2] Faculty of Environmental Science and Engineering, Kunming University of Science and Technology, Kunming 650500, China.
[3] College of Life Sciences, Qufu Normal University, Qufu, Shandong 273165, China.
[4] School of Environmental Science and Engineering, Tianjin University, Tianjin 300072, China.
[5] Helmholtz Centre for Environmental Research– UFZ, Department of Environmental Engineering, Permoserstr. 15, D-04318 Leipzig, Germany.
* Corresponding author: tangjch@nankai.edu.cn

adverse effects on humans and environment. Wastewater is a complex matrix consisting of 99.9% water and 0.1% solids, wherein solids contain organic (70%) as well as inorganic (30%) components (Jain et al. 2021).

Nanotechnology has proven to be an effective and lowest-cost practice for wastewater restoration. Nanomaterials (NMs) and nanoparticles (NPs) are defined as materials and particles with the size (at least one dimension) of their structural components extending from 1 to 100 nm (Buzea et al. 2007). Because of the nanoscale size, the mechanical, electrical, optical and magnetic properties of NMs are distinct from conventional materials of macroscale size (Goswami et al. 2020, Lens et al. 2013). In recent past, research and development of NMs has flourished enormously and NMs have been are effectively utilized in catalytic, pharmaceutical, sensing and biological applications.

It is known that the approach used for synthesizing NMs plays an important part in defining their key properties, stability, morphology, adsorption capacity, catalytic degree, etc. In addition, nanoscale materials have exclusive features, entirely dissimilar from equivalent materials on macroscale. For enhanced operational success, NMs are functionalized with a variety of chemical groups to boost their capacity of degrading the concerned pollutants for wastewater restoration. Meanwhile, the application of NMs in wastewater restoration garnered attention due to their small size, large specific surface area (SSA) and high mobility in aqueous solutions, showing strong adsorption efficiency and reactivity with (pollutant) ions, quick chemical reactions, etc. (Lens et al. 2013). Other significant features of NMs are associated with different behaviors and electron effects. The most vital property of MNs is their large surface-area-to-volume ratio that makes them suitable for wastewater treatment through adsorption (Rahmani et al. 2011), catalysis (Lee et al. 2018), photocatalysis (Eskizeybek et al. 2012), membrane cleaning, disinfection (Hajdu et al. 2012), etc.

Better properties of NMs endorse their feasibility for pollutants degradation and wastewater treatment. Nashaat N Nassar divided nanotechnological wastewater treatment into three core categories: (a) handling and remediation, (b) sensing and detection, and (c) pollution deterrence (Nassar 2013). Multiple NMs have shown satisfactory removal and degradation of heavy metals, organic toxins, inorganic anions and dyes (García-Torres et al. 2017, Salahi et al. 2015, Shirsath and Shirivastava 2015). Smaller particle size, higher surface-area-to-volume ratio and numerous surface active sites encourage chemical reactions and photon absorption by NMs, thus increasing the process efficiency. Superior SSA increases surface states, thus altering the electron and hole activity and affecting reaction dynamics. Broadly studied NMs for wastewater restoration are zero-valent metals (Fe, Ag, Ni and Zn), metal oxides (titanium dioxide (TiO_2), ZnO and Fe oxides), carbon nanotubes (CNTs) and nanocomposites.

Functionalization of NMs offers a promising note for their utilization in environmental remediation. Protection of as-synthesized NMs by casing an appropriate layer of organics/inorganics on their core surface with functional groups improves their functionalities (Anderson et al. 2019). Functionalized NMs supported with nano-adsorbents, nano-catalysts, nano-photocatalysts and nano-membranes have proven superior selectivity and precise sensitivity to remove and eradicate pollutants from surface and sub-surface waters (Sreevidya et al. 2021). Functionalized NMs revealed their potential in getting rid of biological, organic, and inorganic pollutants from wastewater, and commendable literature on the synthesis and applications of functionalized NPs is available (Mohaghegh et al. 2015, Nassar 2013, Wang et al. 2017). With key words 'functionalized nanomaterials for wastewater', 625 published literatures were found until 2022 on the databases of 'Web of Science Core Collection, BIOSIS Previews, MEDLINE®, KCI-Korean Journal Database, Derwent Innovations Index, and SciELO Citation Index'. Figure 1 shows the published literature on functionalized nanomaterials for wastewater restoration from January 2012 to December 2022. Among the total published literature, there were 574 research articles, 74 review articles, 13 proceedings and 9 books.

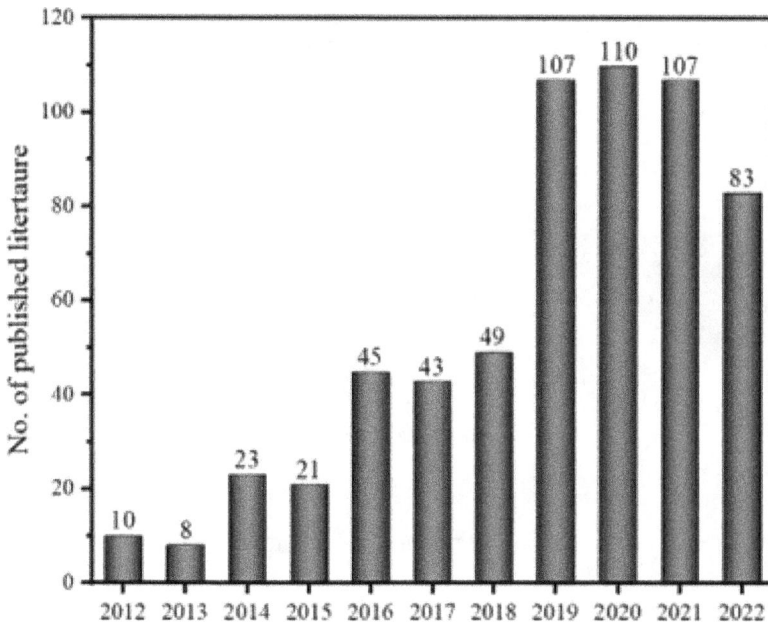

Figure 1: Number of published literature on functionalized nanomaterials for wastewater restoration, from 2012 to 2022.

2. Synthesis of Nanomaterials

There are three core challenges to the mass production of NMs: (a) technical factors of precise nanotechnology use, (b) economic cost-effectiveness for synthesis and operation costs, and (c) impacts on humans and environment.

Two main approaches are practiced to synthesize or structure NMs: 'bottom-up' and 'top-down' (Figure 2) (Lens et al. 2013).

Bottom-up approach: Herein, NMs are synthesized from discrete atoms and molecules linked by chemical bonds, consequently creating larger and more complicated nanostructures through nucleation and growth processes. Chemical synthesis, self-assembly and positional assembly are bottom-up approaches. Chemical synthesis produces particles that are used either directly in the production of bulk materials, or manufactures well-ordered and innovative materials through chemical reactions. Self-assembly involves atoms re-arranged in well-organized nanoscale structures. Positional assembly uses gentle nanoscale tools to manufacture devices at the molecular level; atoms or molecules are consciously manipulated and stationed accordingly (Roy and Bhattacharya 2015).

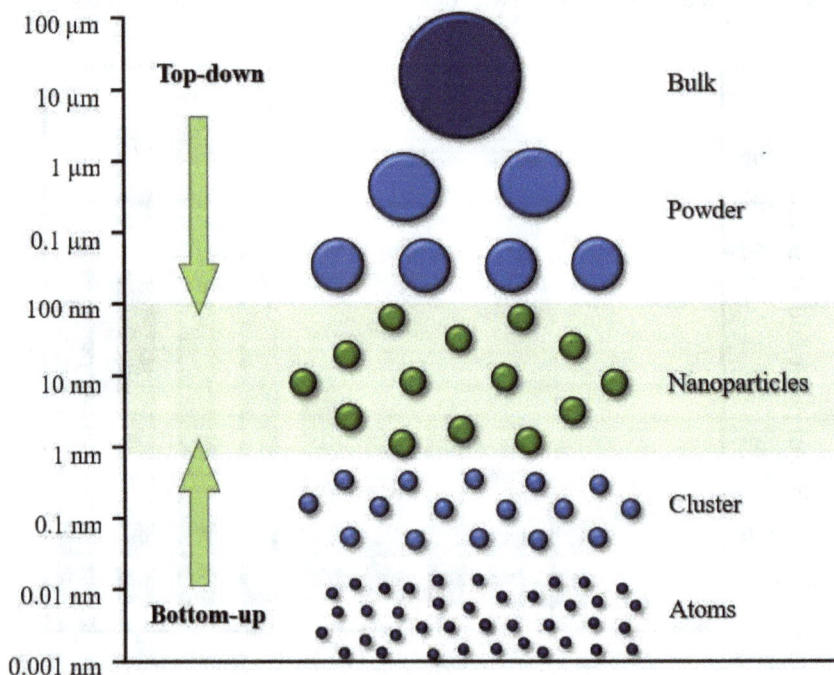

Figure 2: Top-down and bottom-up approaches for the synthesis of NPs (Roy and Bhattacharya 2015, Nassar 2013).

Table 1: Chemical and physical synthesis methods for nanomaterials (Roy and Bhattacharya 2015).

Synthesis Method	Description
Chemical Synthesis Methods	
Microwave irradiation	NM synthesis occurs at low temperature, in a short time, under mild conditions, by direct interaction between microwave irradiation and materials, wherein electromagnetic energy is converted into thermal energy (energy conversion instead of heat transfer). Heat generation from inside the material reduces the reaction time and energy cost. Energy is released as heat due to dielectric losses and molecular friction. Thus, microwave irradiation is quick, modest and competent in energy terms, compared to conventional methods of synthesizing pure NMs with narrow particle size distribution, and the formation of undesired byproducts can be suppressed.
Chemical vapor deposition	The substrate is in contact with one or more volatile sources that react and decompose on the substrate for synthesizing the desired solid materials of superior purity and performance. This method is used in the semiconductor industry to synthesize thin films, and volatile byproducts are eliminated by gas flow over the reaction chamber.
Vapor-phase	NPs are formed in gaseous phase with the condensation of atoms and molecules. Vapor phase synthesis is not new and flame reactors are used for many NP synthesis processes.
Hydrothermal	Single crystal synthesis is dependent on the solubility of minerals in hot water at high pressure in an autoclave. Temperature gradient is maintained at the opposite ends, wherein the warmer end dissolves the nutrients and the cooler end helps in supplementary seed growth. It is the most appropriate method for the development of large high-quality crystals, keeping decent control over their composition.
Micro-emulsion	Synthesis of ultrafine NPs with size < 5 nm and diameter < 50 nm can be done through this method. Particle nucleation rate here is a function of the percolation degree of micro-emulsion droplets. Stabilizer (emulsifier) nature, type and concentration, as well as continuous phase type, reducing agent, surface activity of additives and colloidal stability of micro-emulsion droplets are important for particle size and distribution during NP synthesis with this method.
Sonochemical	NPs are manufactured by irradiating an aqueous or organic dispersion of precursors, with an ultrasonic probe at room temperature. The size of NPs relies on the concentration of the solution and sonication time.

Table 1 contd. ...

...Table 1 contd.

Synthesis Method	Description
Physical Synthesis Methods	
Laser ablation	A high-power laser pulse is used to evaporate substance from the target surface. A supersonic jet of particles is ejected normal to the target surface. Plume, like rocket exhaust, enlarges away from the target, with a robust forth-directed velocity distribution of particles. Ablated species condense on the substrate positioned against the target. Ablation happens in a vacuum chamber (vacuum or gas (oxygen)).
Sputtering	Thin material films are deposited on to the substrate surface. First, a gaseous plasma is created, which accelerates ions into the target material that is eroded by incoming ions via energy transfer and ejected as neutral particles (distinct atoms or atoms/molecules clusters). Ejected particles travel in a straight line unless they encounter other particles or surfaces.
Spray route pyrolysis	A modest and quick aerosol decomposition (spray pyrolysis) process is used for the constant synthesis of NPs with modifiable sizes, narrow size distribution, and better crystallinity and stoichiometry.
Inert gas condensation	It is used to prepare metallic Fe NPs by evaporation and aggregation in an inert gas atmosphere. This most advanced (but expensive) method is considered to be a controlled method of synthesizing NPs with exact shape and size. The formed NPs immediately strike the low-pressure inert gas and consequently smaller and adjustable NPs are synthesized.

Top-down approach: NMs are synthesized from bulk entities with no configuration control at the atomic or nano level. It includes various physical synthesis (and breaking) methods, like milling, grinding, etching, laser radiation, etc. Precision engineering and lithography are top-down approaches. Precision engineering is deployed in the production of semiconductors used as substrates for computer chips, and for printing patterns on ceramic wafers. Lithography involves pattern creation through contact with light, ions or electrons, and successive etching or deposition of material on the surface, with possible accuracy of 10 nm or less (Roy and Bhattacharya 2015).

Between chemical and physical synthesis of NMs, chemical synthesis is favored due to the sensitive shape, size and composition of NMs with novel properties, and its potential for cost-effective large-scale NMs synthesis (Roy and Bhattacharya 2015). In addition, brief details of a few other chemical and physical synthesis methods of NMs are listed in Table 1.

3. Functionalities of Nanomaterials

Functionalization is defined as adding polymeric and organic moieties on to the surface of NMs. Several organic molecules and functional groups are connected to NPs through physical or chemical approaches to expand the features and structures of NMs. Functionalized NMs got attention due to their numerous applications in biomedicine, bioenergy and biosensor fields (Hussain 2020). Simple NMs may possess higher agglomeration and low surface reactivity and mobilization. Such limits are overcome by functionalization that assists in the development of stable and mobile NMs with small band gap and improved surface reactivity. NPs are functionalized with several chemical groups to improve their interaction with pollutants. Functionalization contributes to oxygen, nitrogen, sulfur, phosphorus or other functional groups on the surface of NMs, develops their dispersion and thus increases SSA (Sreevidya et al. 2021). For a brief discussion, carboxylic, amine, phosphonic and sulfonic functional groups, along with their respective formation mechanisms, have been mentioned in Figure 3.

Moreover, synthesis and application of polymer-functionalized NMs have flourished over the years, and the general functional moieties are N-donors (amides, amines) and O-donors (ethers, alcohols), both proving themselves to be of great interest. Advanced functionalized NMs own superior chemical and antimicrobial active properties and are also recognized for their improved capability for pollutant removal from environment media (Hussain 2020, Mahajan et al. 2022). Functionalization of NMs is normally realized by direct and indirect functionalization

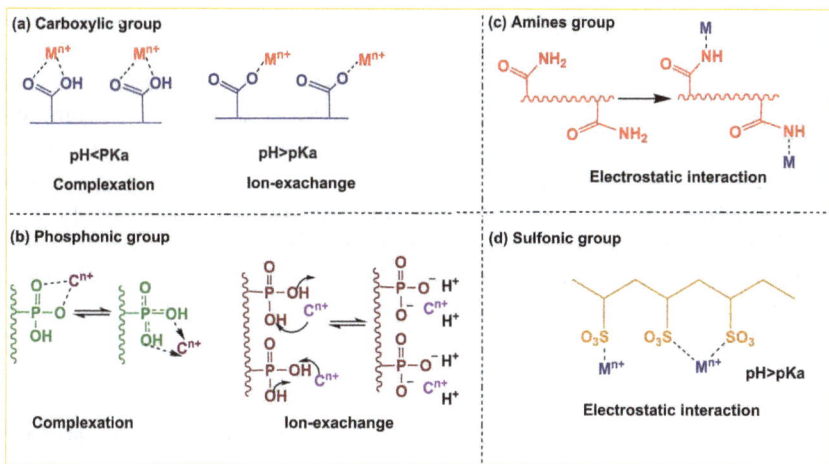

Figure 3: Illustration of dominant mechanisms on carboxylic, amine, phosphonic and sulfonic functional groups (Makvandi et al. 2021).

methods. Direct (co-condensation and in situ) functionalization includes covalent, non-covalent, inorganic, heteroatom doping, and immobilization-mediated functionalization methods, while indirect functionalization consists of grafting.

3.1 Direct Functionalization

NMs are modified through complex and switchable bifunctional units via direct functionalization by adding –COOH, –OH and –NH$_2$ functional groups. Functionalized NMs offer several merits with the formation of uniform surfaces, improved tuning of integrated units, and the ability to utilize different functional units.

3.1.1 Covalent Functionalization

In covalent functionalization, modifying groups are attached to NPs, mediated by organic coupling interactions to establish stable bonds. Functional groups are connected to the NP surface, which bind further with selected functional materials through diverse organic reactions. This method is generally used in the synthesis of nano Fe^0 and carbon-based (C-based) NMs. In this method, NMs are obtained by oxidation, halogenation, amidation, thiolation, hydrogenation, sulfonation, doping, alkali activation, etc. Compared to non-covalent bonds, covalent bonds are stronger and resist separation, regardless of the process involved (Mahajan et al. 2022). However, damage to the structure and inherent properties could be the demerits of covalent functionalization.

3.1.2 Non-covalent Functionalization

Non-covalent functionalization is a simple process and provides improved structural and electronic properties, with minimum damage to NMs. Non-covalent hydrophobic linkages and relationships between amphiphilic units and aromatic groups of C-based NMs have been studied in detail. In addition, they are consumed to decrease the hydrophobic interface, thereby increasing the solubility of C-based NMs under freezing conditions (Mahajan et al. 2022). However, weak coating stability could be a demerit of non-covalent functionalization.

3.1.3 Inorganic Functionalization

Sigle metals and a combination of metals are used for the surface modification of NMs through blending and in situ synthesis. The former includes mixing solutions and molten compounds, whereas the latter involves the combination of precursors and C-based materials, followed by the manufacture of NPs on the surface of C-based NMs (Mahajan et al. 2022).

3.1.4 Functionalization by Heteroatom Doping

A variety of functionalized NMs are synthesized with heteroatom (N, O, B and P) doping. Doping is generally accomplished through the hydrolysis of precursors or ion implantation, and the resultant NMs possess enhanced hydrophilicity and electrical conductivity (Mahajan et al. 2022). Among doped elements, N has been studied most widely, and it comes from ammonia, pure organic amines, and amines/ammonia sources, while hydrocarbons serve as a carbon source to build NMs. Among Ba, Sn, Fe, Co and Al metals, Fe is mostly used with non-metal supports such as N, F, S or C, or with other metals like Pd, Au, Ag and Cu, and transition metals; the resultant NMs thus provide enhanced reactivity and adsorption properties. Differences between the band gaps of different metallic and mineral compositions have been observed, and the band gap of porous structures is lower than that of solid structures. NMs functionalized by doping favor the generation of reactive free radicals to degrade environmental pollutants (Wang et al. 2017).

3.1.5 Functionalization by Immobilization

Immobilization mainly allows for improving the active sites of NMs, offering superior SSA and stability. Immobilization is usually performed on porous substrates such as green tea, kaolin, bentonite, carbon, silica etc. and oxides of Zn and Fe are usually immobilized on enzymes (Mahajan et al. 2022).

3.2 Indirect or Post-synthetic Functionalization (Grafting)

Indirect functionalization is accomplished in two steps for grafting bifunctional units on to the surface of NMs. First, the linking group interaction occurs, followed by the functional group's transition to the conjugation site. Silylation is the grafting process of integrating a silyl group with a NM, and hybrid NMs can be fabricated using functional silica supports. Grafting is usually carried out using oxygen-containing molecules (–COOH groups) already present in C-based NMs; they react with thionyl chloride ($SOCl_2$) and lead to acyl chloride formation. Thus, C-based NMs functionalized with acyl chlorides can form C-based

NMs with ester functional groups after reaction with alcohols (Mahajan et al. 2022). The surface of C-based NMs can be decorated with polymers through covalent as well as non-covalent bonds. Polymer coatings expand many features of C-based NMs, such as consistency, stiffness, electrical conductivity and adsorption capacity. The –COOH and –OH groups of C-based NMs contribute to stability during doping with a polymer, and this functionality can be easily obtained through covalent bonding (Czerw et al. 2001).

4. Characterization of Nanomaterials

Synthesis and characterization of NMs are of great importance in nanoscience and nanotechnology in order to learn about the physical characteristics of materials for understanding their consequential properties. Characterization studies are principally vital for materials with nanoscale length as considerable variations in material properties occur with dimensional modifications. Brief details of characterization of NMs are itemized in Table 2.

Table 2: Characterization of nanomaterials.

Characterization	Description
X-ray diffraction	To provide accurate information on the composition, crystal structure and crystalline grain size.
X-ray photoelectron spectroscopy	To quantitatively examine the surface chemical composition and the local chemical environment of the given atoms.
Scanning electron microscopy	To analyze the surface morphology (texture and chemical composition).
Energy dispersive X-ray spectroscopy	To identify and quantify the elements (elemental composition).
Transmission electron microscopy	To reveal the precise structural details, size distribution and morphology.
Brunauer-Emmett-Teller	To measure the specific surface area.
Fourier transform infrared spectroscopy	To identify (surface) functional group modification(s).
Magnetization measurement	To determine the magnetic moment magnitude (magnetic moment per unit volume indicates magnetization).
Zeta potential	To calculate the mobility of particles in an electric field.
Atomic absorption spectrometry	To quantitatively determine the elements through the absorption of radiation by free gaseous atoms.

5. Application of Functionalized Nanomaterials for Wastewater Restoration

5.1 Nanomaterials for Adsorption

NMs possess two vital features that make them brilliant adsorbents: innate large SSA, and surface functionalities or capacity to chemically react with neighboring pollutants. Physical, chemical and structural features are also linked to their exterior surface, apparent size and interior composition. These features make NPs effective adsorbents for multiple pollutants in wastewater and keep them stable for a long time, also resulting in adsorbent degradation and expanding adsorption efficiency. Nano-adsorbents are categorized into many groups such as metallic nano-particles, nanostructured mixed oxides, magnetic NPs, metallic oxide NPs and carbonaceous NMs (CNTs, NPs and nanosheets) (Table 3). Sumio Iijima discovered CNTs in 1991; since then, CNTs have made their way into applications like adsorption of heavy metals and organic pollutants– due to the large SSA and superior structure of NMs, and the arrangement of carbon atoms. Exceptional mechanical, electrical, chemical and optical properties, even better than activated carbon and other carbon materials, make CNTs a better adsorbent and hence they are called the 'material of the 21st century' (Ren et al. 2011).

Metal-based NPs also have better size, surface chemistry, aggregation effect, shape and fractal dimension, chemical composition, crystallinity and solubility characteristics. The widely used metal oxide adsorbents for the adsorption of pollutants in wastewater with high BET SSA, least environmental impacts, low solubility and no secondary pollutants are iron oxides (Fe_xO_y) and oxides of silicon, titanium, zinc, magnesium, manganese and tungsten. The surface functionalities of metal and metal oxide NPs are reformed by water pH as their surface endures protonation or deprotonation, which depend on water pH and NPs surface charge (Nassar 2013). Adsorption reaction is based on the electrostatic interaction of dissolved ions in wastewater and NM surface; hence, a change in the solution's pH considerably affects the interaction strength. The surface of the nanocomposite is to be acidic– having a positive charge– to attract anions, and basic – having a negative charge– to attract cations (Qu et al. 2013). Adsorption of heavy metals and organic pollutants by nano-adsorbents depends immensely on electrostatic interactions between the nano-adsorbent's surface and ions. The surface charge of nano-adsorbents changes with pH, and their interaction with pollutants is thus affected. With this property, adsorptive efficacy of nano-adsorbents is boosted for organic compounds; hence, nano-adsorbents surface-modified with surfactants facilitate the removal of complex compounds and organic pollutants (Zhao et al. 2008).

5.2 Nanomaterials for Catalysis

The use of nano-catalysts in catalytic reduction and chemical oxidation is advantageous over traditional treatment approaches as they target recalcitrant pollutants, have low reaction times and are capable of altering wastes into valued byproducts (Table 3). C-based NMs, if utilized with metals or metal oxides (i.e., ZnO and graphene), are effective for the removal of lethal pollutants from wastewater. This is because constructive alterations occur due to the synergistic effect of C-based NMs and metallic NPs. Metal, bimetallic and metal oxide NPs have demonstrated themselves as excellent nano-catalysts in catalytic reduction and oxidation reactions due to their smaller particle size (large surface-area-to-volume ratio), superior activity associated with the nanoscale size, and surface functionality. They show robust catalytic activity through which pollutants are oxidized into non-toxic compounds or are transformed into environmentally tolerable ultimate products through advanced oxidation (Lens et al. 2013). Advanced oxidation is based on the formation of very reactive radicals that interact directly with pollutants. During catalytic reduction, nano-catalysts convert refractory pollutants into less toxic or non-toxic compounds (Wenying et al. 2008).

Zinc (Zn) NPs possess a higher negative standard reduction potential that makes them qualified for the removal of various redox-labile pollutants. Due to this trait, the pollutant degradation rate of Zn NPs is expected to be faster than nano-zero-valent Fe (nano-Fe0) (Bokare et al. 2013). Iron oxide (i.e., magnetic magnetite (Fe_3O_4) and maghemite (γ-Fe_2O_4), and nonmagnetic hematite (α-Fe_2O_3)) NPs used for removing heavy metals and organic pollutants are more beneficial due to their simplicity and accessibility. Nano-Fe0 and nano-Fe0 supported on carbon are broadly studied metal NPs in wastewater restoration through catalytic reduction and advanced oxidation (Ahmad et al. 2021b). Among NMs, only nano-Fe0 has touched commercial application; other technologies are still being discovered. For instance, nano-Al0 is thermodynamically unstable due to its high reductive ability in aqueous solutions, which results in oxide/hydroxide formation on the surface, thus obstructing electron transfer from the metal surface to the pollutants (Rivero-Huguet and Marshall 2009). For commercial application, Pd offers selective removal of pollutants (chlorinated hydrocarbons) from wastewater, and Pd-based nano-catalyst (Pd/Fe_3O_4) has shown superior hydro dechlorination and recovery via magnetic separation from wastewater (Hildebrand et al. 2009).

5.3 Nanomaterials for Photocatalysis

For wastewater restoration, photocatalysis is a suitable technique for removing environmental pollutants (Table 3). Nano-photocatalysts

Table 3: Performance of various types of functionalized nanomaterials for wastewater restoration.

Technology	Nanomaterial	Pollutant	Conditions	Performance	Remarks	Reference
Adsorption	Nano-Fe^0	As(III)	pH: 7, T: 10 min, dose: 1 g/L	99.9%	Nano-Fe0 showed brilliant removal and adsorption was better than centrifugation or filtration.	(Rahmani et al. 2011)
	Magnetic nano-adsorbent	Zn(II)	pH: 5.5, T: 1.5 h, dose: 2.5 g/L	95%	Langmuir and Freundlich models; adsorption was spontaneous and endothermic	(Shirsath and Shrivastava 2015)
	Green Cu NPs	Ibu, Nap, Dic	[Conc]0: 10-40 mg/L, pH: 4.5, dose: 10 mg, t: 25°C, T: 1 h	33.9, 33.9 and 36.0 mg/g	Langmuir; adsorption was spontaneous, endothermic and physical	(Husein et al. 2019)
	As-grown and graphitized CNTs	DCB	[DCB]0: 20 mg/l, pH: 5.5, T: 24 h	30.8 and 28.7 mg/g	Freundlich; adsorption reaction was highly spontaneous and endothermic	(Peng et al. 2003)
Catalysis	Fe0/Pt	MB	Dose: 10 mg, [MB]0: 30 mg/L, pH: 3, H2O2: 5%, T: 1 h	Complete oxidative degradation	Fenton reaction	(Lee et al. 2018)
	Polystyrene@ZIF-Zn-Fe	RhB	[RhB]: 20 mg/L, H2O2: 12%; micromotor conc.: 0.5 g/L, t: 25°C	47% reduction of RhB in 2.5 h	Catalytic oxidation	(Wang et al. 2017)

Table 3 contd. ...

...Table 3 contd.

Technology	Nanomaterial	Pollutant	Conditions	Performance	Remarks	Reference
	CoNi@Pt nanorods	4-NP, MB and RhB	4-NP: 0.1 mM, MB: 0.06 mM, RhB: 0.05 mM, NaBH4: 25 mM, catalyst dispersion: 1 mg/mL	Complete removal	Catalytic degradation	(García-Torres et al. 2017)
Photocatalysis	Polyaniline/ZnO	MB and MG	Visible region (UV and natural sunlight), dose: 0.4 g/L, T: 5 h	99%	Good photocatalytic stability; reused 5 times with minor gradual loss of activity	(Eskizeybek et al. 2012)
	ZnO/Zn	MB and CR	365 nm (UV), pH: 6–8, T: 4.5 h	98 and 92%	ZnO/Zn nanocomposite using corn starch and cellulose as chelating agents for photodegradation	(Lin et al. 2014)
	Carbon nanodots-TiO$_2$	2,4-DCP	≥ 420 nm (vis), [DCP]$_0$: 25 mM, T: 8 h, dose: 0.2 g/L	96%	Carbon nanodots increased the photocatalytic response of TiO2 in the visible range.	(Ortega-Liébana et al. 2016)
	Ag3PO4/BiPO4/Cu(tpa).graphene	Atrazine	≥ 420 nm (vis) T: 2 h		Metal-organic framework–graphene-based nanocomposites showed 80% degradation.	(Mohaghegh et al. 2015)
Membrane and disinfection	Biodegradable poly-gamma-glutamic acid (80–350 nm)	Pb(II)	Conc.: 1 mg/mL, pH: 6	99.8%	Ultrafiltration for efficient and fast Pb(II) removal	(Hajdu et al. 2012)

	Nano-porous membrane filtration	Oil, grease and BOD	t: 45°C, flow: 1.3 m/s, P: 4 bar, salt conc.: 11.2 g/L, pH: 10	Oil (99%), grease (80%), BOD (76%)	Membranes had asymmetric structure.	(Salahi et al. 2015)
	Carbon nanofiber membrane (126–554 nm)	Metal and metal oxide NPs	t: 30°C, pH: 7	95%	Mechanically strong and bendable at high pressure or in vacuum	(Faccini et al. 2015)
	Ag and ZnO NPs	*E. coli*	pH: 6.4–8.4, T: 60 min, storage: 24 h	3.18 (log removal value)	Fairly robust to changing pH	(Venis and Basu 2021)

Legend: Ibuprofen (Ibu), Naproxen (Nap), Diclofenac (Dic) Dichlorobenzene (DCB), 4-Nitrophenol (4-NP), Methylene blue (MB), rhodamine B (RhB), Malachite green (MG), Congo red (CR), 2,4-dichlorophenol (2,4-DCP).

consist of semiconductor metals that can degrade diverse refractory organic pollutants. Photocatalysis comprises three sequential stages: photo excitation, charge separation and electron-hole production. It is a noteworthy advanced oxidation method using ultraviolet (UV) light as an energy source. UV light irradiation produces very reactive hydroxyl radicals ($^\bullet$OH) for the remediation of recalcitrant compounds in wastewater. Compared to Fenton, photon-Fenton and H_2O_2 catalytic reactions, photocatalysis has a better potential as it operates under ambient conditions and uses low-cost non-toxic photocatalysts, and atmospheric air as the oxidant.

In recent years, photocatalysis has been successfully applied to wastewater restoration; therein, a combination of light and catalyst steadily oxidizes the pollutants into low molecular weight intermediates and products and then ultimately transforms them into CO_2, H_2O and anions (i.e., NO_3^-, PO_4^{3-} and Cl^-). Photocatalysis is about the interaction of light energy with nano-photocatalysts (oxide or sulfide semiconductors). TiO_2 and ZnO NPs have been widely explored recently due to their high photocatalytic activity, economy, photostability, and chemical and biological stability (Daneshvar et al. 2004, Rawal et al. 2013). TiO_2 has a large band gap (3.2 eV) that involves UV excitation at wavelengths < 390 nm to encourage charge separation within particles, and the synergistic effect of TiO_2 and UV generates reactive free radicals to degrade heavy metals and persistent organic pollutants in water (Guo et al. 2015, Moon et al. 2014). ZnO NPs have a wide band gap in near-UV spectrum, higher oxidation ability and superior photocatalytic activity; they are also environment-friendly due to their compatibility with organisms (Schmidt-Mende and MacManus-Driscoll 2007). The large band gaps of TiO_2 and ZnO could be a drawback as these NPs work better in UV excitation and their performance is quite ordinary under visible light. Further research is needed to expand the photocatalytic properties of TiO_2 and ZnO NPs in visible light and UV. Cadmium sulfide (CdS) is also an eminent semiconductor with a 2.42 eV band gap and operates at wavelengths < 495 nm; it has been thoroughly applied as a photocatalyst to treat industrial wastewater containing dyes (Tristão et al. 2006). The mechanism of photocatalysis is based on the photoexcitation of electron in nano-photocatalysts. Irradiation with light (UV for TiO_2) produces holes (h^+) and exited electrons (e^-) in their conduction band. In water, the h^+ is trapped by H_2O molecules to produce hydroxyl radicals ($^\bullet$OH, an unselective but powerful agent) which oxidize and degrade refractory pollutants into water and gaseous products (Anjum et al. 2019).

5.4 Nanomaterials for Membranes and Water Disinfection

Membrane filtration designed using NMs is one of most practical approaches that comes with superior functionalities, i.e., catalytic reactivity, permeability, and fouling resistance. It offers benefits in terms of water quality restoration, efficient disinfection, and low area requirement for operation; it is also extremely cost-effective, competent, and easy to design (Jang et al. 2015). Nano-membranes are of two types: mixed matrix nanocomposite membranes and thin-film nanocomposite membranes (Ahmad et al. 2021a). The former refers to NPs that disperse into the polymer casting solution before membrane casting, known as NP blend membranes. The latter refers to NPs that are embedded within a thin film layer or are self-assembled on the membrane surface by dispersion and dip-coating, or pressurized deposition on the synthesized membrane.

Incorporation of NPs in the membrane structure assists the expansion of the surface nanostructure and improves adhesion, which further improves the latter's 'molecular sieve' property, salt rejection selectivity, activity for pollutant interaction, and surface hydrophilicity to hinder the attachment of microorganisms by deactivating them (El Saliby et al. 2009). Polysulfone membrane impregnated with Ag NPs has shown effective efficiency against *Escherichia coli* K12 and *Pseudomonas mendocina*, proving itself to be competent for virus removal and biofouling resistance due to the release of Ag^+ ions that bar the attachment of bacteria to the membranes (Zodrow et al. 2009).

Water disinfection refers to the use of chemicals (disinfectants) to inactivate and eliminate microorganisms on inert surfaces (Table 3). Disinfectants affect and damage cell wall, cell permeability, protoplasm and enzymatic activity, consequently upsetting the activity of microorganisms that are incapable of reproduction, and causing their death (Ahmad et al. 2021a). However, the by-products formed during water disinfection have proved to be a problem with traditional disinfectants, like chlorine, chloramines and ozone, that react with wastewater substances to form the by-products. NMs, specifically inorganic antibacterial NMs, can eliminate this drawback as their reaction suppresses the formation of by-products (Ahmad et al. 2021a). Various NMs (including chitosan, Ag NPs, TiO_2, fullerene NPs, CNTs, etc.) have been established to have great antimicrobial activity. These NMs are mild oxidants; they are inert in water and can generate some harmful by-products as well. NMs in water disinfection can perform multiple roles: direct action on (bacterial) cells to stop electron passage, breaking through the cell membrane, oxidation of cellular components and hydroxyl radicals (acting as photocatalysts), and establishment of dissolved metal ions to harm the cellular components (Li et al. 2008).

ZnO NPs as a nano-disinfectant for water treatment have demonstrated the release of reactive free radicals and Zn^{2+}; on direct contact with cell membrane, both – the reactive radicals and Zn^{2+} ions– infiltrate the intracellular contents of cell walls and cause fatal damage to the microorganisms, and the dented and damaged cells have been attributed to ZnO (Dimapilis et al. 2018). Silver NPs, generally used as nano-disinfectants, are very lethal to microorganisms and have shown robust antibacterial and antimicrobial effects on viruses, bacteria and fungi (Lu et al. 2016). The antibacterial activity of Ag NPs has been explored for *Staphylococcus* as positive and *E. coli* as negative bacteria, wherein Ag began protein leakage and lactate dehydrogenase inactivation, converting lactate into pyruvate. Besides, the high SSA of Ag NPs helps release Ag^+ to oxidize the microbes associated with reactive free radical generation (Kim et al. 2011).

6. Cost of Functionalized Nanomaterials

The cost of functionalized NMs needs to be well-known and recognized. The cost of nanofibers membranes is about $20€/m^2$, which is less expensive than commercial membranes ($50€/m^2$) (Bjorge et al. 2009). The cost of synthesis of nano-membranes is about $5€/m^2$, with the main cost (75%) driven by nonwoven support, showing that cheaper supports can decrease the total cost to less than that of conventional polymeric membranes ($14–50€/m^2$) (Fatarella et al. 2009). The price of single- and multi-wall CNTs is about 1500\$/g and 900\$/g respectively; the price of graphitized nanofibers with > 95% trace metal basis and graphite platelet nanofibers with > 98% carbon basis is about 80\$/g and 250\$/g respectively (http://www.sigmaaldrich.com). Furthermore, the cost of raw materials for NMs can vary with the synthesis method and support material (Pervez et al. 2020), and optimum parameters for pollutant removal can provide a sustainable route for wastewater restoration.

7. Environmental Impact of Functionalized Nanomaterials

In the past couple of decades, various NMs have made their place in the commercial market for direct as well as indirect applications ranging from environmental decontamination to health. The application of NMs should not be focused in just environmental and health sectors; rather, their role and effects on environment and humans throughout their use and lifecycle must be evaluated. Yet, little is known about the impact of synthetic NM exposure on humans and environment (Nel et al. 2006). The available literature points out some adverse effects of NMs on the environment (Nel et al. 2006), whereas another study says that various NM applications pose no new health or safety risks (Nassar 2013). The use of NMs in wastewater restoration is still limited to an early stage. Nonetheless, potential complications can arise in vitro and in vivo applications of functionalized NMs. Uncertainties associated with functionalized NMs for health and environment need to be considered before their widespread application.

In terms of maintenance, a probable problem is the elimination of accumulated pollutants from NMs and the possibility to reuse MMs. Further research is required to minimize the demerits of nanotechnology and utilize its whole potential in wastewater restoration. Environmental hazards are restricted to a few NPs as they are more reactive and mobile, not combined with any support materials, could have negative influences due to their smaller size or chemical properties (Pidgeon et al. 2004). The use of such NMs (e.g., nanopowder) must be stopped, and we need to make some regulatory policies regarding their environmental impacts (Pidgeon et al. 2004).

8. Conclusion and Future Prospects

This chapter provides an overview of the synthesis of functionalized NMs through different synthesis techniques and functionalities of NMs via direct and indirect functionalization. Direct functionalization includes covalent, non-covalent, inorganic and heteroatom doping, and immobilization, while indirect functionalization includes grafting and post-synthetic techniques. Various characterization techniques of functionalities on NMs and the latter's role in wastewater restoration has been discussed for adsorption, catalysis, photocatalysis, membranes and water disinfection; the cost and environmental impacts of functionalized NMs have also been assessed for progressive wastewater restoration. Comprehensive research has been conducted for the development of NPs for use in wastewater restoration, and a few prospects have been suggested herein.

NPs are efficiently manufactured using the green synthesis approach, showing suppressed agglomeration and selective reactivity. Further efforts are required to commercialize functionalized NM synthesis through green means.

Membrane fouling is the main demerit of NMs. Reversible and irreversible fouling result due to the contact between wastewater components and membranes, wherein reversible fouling has a higher adverse fouling rate than irreversible fouling. As a solution, reversible fouling is minimized with physical actions (air sparging and backflushing) and irreversible fouling is mitigated through chemical treatment.

Though nanotechnology offers plenty of enhancements in wastewater restoration, it can still have a bad impact on the organisms and humans exposed to then. Therefore, a lot of research is required to offer effective conclusions regarding their possible risks to humans and environment.

There is a lot of discharge of NMs into the environment due to nanotechnology and industrial growth, the research on risk assessment and management of NMs is necessary to expand to fill the knowledge gaps in this research direction.

Effective education in nanotechnology use for sustainability will be an important step towards learning experiences to encourage the services and skills needed to bring about sustainable change without toxic environmental impacts. Furthermore, life cycle assessment of NMs is vital to deliberate on their overall merits and demerits.

Till date, most NMs are not cost-competitive in comparison with conventional materials. Additional studies are required for cost-effective synthesis of NMs, testing their efficiency at a large scale, and finding applications where only a small amount of NMs is required.

Acknowledgments

This study was supported by Natural Science Foundation of Tianjin (20JCZDJC00700), National Natural Science Foundation of China (U1806216, 41877372), International Postdoctoral Exchange Fellowship Program of China supported by Tianjin Municipality, National Key R&D Program of China (2018YFC1802002), and 111 program, Ministry of Education, China (T2017002).

References

Ahmad, N., Goh, P., Zulhairun, A., Wong, T. and Ismail, A. 2021a. The Role of Functional Nanomaterials for Wastewater Remediation.

Ahmad, S., Liu, X., Tang, J. and Zhang, S. 2021b. Biochar-supported nanosized zero-valent iron (nZVI/BC) composites for removal of nitro and chlorinated contaminants. Chemical Engineering Journal 133187.

Anderson, S.D., Gwenin, V.V. and Gwenin, C.D. 2019. Magnetic functionalized nanoparticles for biomedical, drug delivery and imaging applications. Nanoscale Research Letters 14(1): 1–16.

Anjum, M., Miandad, R., Waqas, M., Gehany, F. and Barakat, M. 2019. Remediation of wastewater using various nano-materials. Arabian Journal of Chemistry 12(8): 4897–4919.

Bjorge, D., Daels, N., De Vrieze, S., Dejans, P., Van Camp, T., Audenaert, W. et al. 2009. Performance assessment of electrospun nanofibers for filter applications. Desalination 249(3): 942–948.

Bokare, V., Jung, J.-l., Chang, Y.-Y. and Chang, Y.-S. 2013. Reductive dechlorination of octachlorodibenzo-p-dioxin by nanosized zero-valent zinc: Modeling of rate kinetics and congener profile. Journal of Hazardous Materials 250: 397–402.

Buzea, C., Pacheco, I.I. and Robbie, K. 2007. Nanomaterials and nanoparticles: Sources and toxicity. Biointerphases 2(4): MR17–MR71.

Czerw, R., Guo, Z., Ajayan, P.M., Sun, Y.-P. and Carroll, D.L. 2001. Organization of polymers onto carbon nanotubes: A route to nanoscale assembly. Nano Letters 1(8): 423–427.

Daneshvar, N., Salari, D. and Khataee, A. 2004. Photocatalytic degradation of azo dye acid red 14 in water on ZnO as an alternative catalyst to TiO2. Journal of Photochemistry and Photobiology A: Chemistry 162(2-3): 317–322.

Dimapilis, E.A.S., Hsu, C.-S., Mendoza, R.M.O. and Lu, M.-C. 2018. Zinc oxide nanoparticles for water disinfection. Sustainable Environment Research 28(2): 47–56.

El Saliby, I., Shon, H., Kandasamy, J. and Vigneswaran, S. 2009. Nanotechnology for water and wastewater treatment: in brief. Water and Wastewater Treatment Technologies. Encyclopedia of Life Support Systems (EOLSS).

Eskizeybek, V., Sarı, F., Gülce, H., Gülce, A. and Avcı, A. 2012. Preparation of the new polyaniline/ZnO nanocomposite and its photocatalytic activity for degradation of methylene blue and malachite green dyes under UV and natural sun lights irradiations. Applied Catalysis B: Environmental, 119: 197–206.

Faccini, M., Borja, G., Boerrigter, M., Morillo Martín, D., Martìnez Crespiera, S., Vázquez-Campos, S. et al. 2015. Electrospun carbon nanofiber membranes for filtration of nanoparticles from water. Journal of Nanomaterials, 2015.

Fatarella, E., Iversen, V., Grinwis, S. and Paulussen, S. 2009. Textile material for membrane filtration. AMADEUS Final Report.

García-Torres, J., Serrà, A., Tierno, P., Alcobé, X. and Vallés, E. 2017. Magnetic propulsion of recyclable catalytic nanocleaners for pollutant degradation. ACS Applied Materials & Interfaces 9(28): 23859–23868.

Goswami, A., Kadam, R.G., Tuček, J., Sofer, Z., Bouša, D., Varma, R.S. et al. 2020. Fe (0)-embedded thermally reduced graphene oxide as efficient nanocatalyst for reduction of nitro compounds to amines. Chemical Engineering Journal 382: 122469.

Guo, M., Song, W., Wang, T., Li, Y., Wang, X. and Du, X. 2015. Phenyl-functionalization of titanium dioxide-nanosheets coating fabricated on a titanium wire for selective solid-phase microextraction of polycyclic aromatic hydrocarbons from environment water samples. Talanta 144: 998–1006.

Hajdu, I., Bodnár, M., Csikós, Z., Wei, S., Daróczi, L., Kovács, B. et al. 2012. Combined nano-membrane technology for removal of lead ions. Journal of Membrane Science, 409: 44–53.

Helmer, R. and Hespanhol, I. 1997. Water Pollution Control: A Guide to the Use of Water Quality Management Principles. CRC Press.

Hildebrand, H., Mackenzie, K. and Kopinke, F.-D. 2009. Pd/Fe3O4 nano-catalysts for selective dehalogenation in wastewater treatment processes — Influence of water constituents. Applied Catalysis B: Environmental 91(1-2): 389–396.

Husein, D.Z., Hassanien, R. and Al-Hakkani, M.F. 2019. Green-synthesized copper nano-adsorbent for the removal of pharmaceutical pollutants from real wastewater samples. Heliyon 5(8): e02339.

Hussain, C.M. 2020. Handbook of Functionalized Nanomaterials for Industrial Applications. Elsevier.

Jain, K., Patel, A.S., Pardhi, V.P. and Flora, S.J.S. 2021. Nanotechnology in wastewater management: a new paradigm towards wastewater treatment. Molecules 26(6): 1797.

Jang, J.H., Lee, J., Jung, S.-Y., Choi, D.-C., Won, Y.-J., Ahn, K.H. et al. 2015. Correlation between particle deposition and the size ratio of particles to patterns in nano-and micro-patterned membrane filtration systems. Separation and Purification Technology 156: 608–616.

Kim, S.-H., Lee, H.-S., Ryu, D.-S., Choi, S.-J. and Lee, D.-S. 2011. Antibacterial activity of silver-nanoparticles against Staphylococcus aureus and Escherichia coli. Microbiology and Biotechnology Letters 39(1): 77–85.

Lee, C.-S., Gong, J., Oh, D.-S., Jeon, J.-R. and Chang, Y.-S. 2018. Zerovalent-iron/platinum Janus micromotors with spatially separated functionalities for efficient water decontamination. ACS Applied Nano Materials 1(2): 768–776.

Lens, P., Virkutyte, J., Jegatheesan, V. and Al-Abed, S. 2013. Nanotechnology for Water and Wastewater Treatment. Iwa Publishing.

Li, Q., Mahendra, S., Lyon, D.Y., Brunet, L., Liga, M.V., Li, D. et al. 2008. Antimicrobial nanomaterials for water disinfection and microbial control: Potential applications and implications. Water Research 42(18): 4591–4602.

Lin, S.-T., Thirumavalavan, M., Jiang, T.-Y. and Lee, J.-F. 2014. Synthesis of ZnO/Zn nano photocatalyst using modified polysaccharides for photodegradation of dyes. Carbohydrate Polymers 105: 1–9.

Lu, H., Wang, J., Stoller, M., Wang, T., Bao, Y. and Hao, H. 2016. An overview of nanomaterials for water and wastewater treatment. Advances in Materials Science and Engineering 2016.

Mahajan, G., Kaur, M. and Gupta, R. 2022. Green functionalized nanomaterials: Fundamentals and future opportunities. *In*: Green Functionalized Nanomaterials for Environmental Applications, Elsevier, pp. 21–41.

Makvandi, P., Iftekhar, S., Pizzetti, F., Zarepour, A., Zare, E.N., Ashrafizadeh, M. et al. 2021. Functionalization of polymers and nanomaterials for water treatment, food packaging, textile and biomedical applications: A review. Environmental Chemistry Letters 19(1): 583–611.

Mohaghegh, N., Tasviri, M., Rahimi, E. and Gholami, M. 2015. Comparative studies on Ag3PO4/BiPO4–metal-organic framework–graphene-based nanocomposites for photocatalysis application. Applied Surface Science 351: 216–224.

Moon, G.-h., Kim, D.-h., Kim, H.-i., Bokare, A.D. and Choi, W. 2014. Platinum-like behavior of reduced graphene oxide as a cocatalyst on TiO2 for the efficient photocatalytic oxidation of arsenite. Environmental Science & Technology Letters 1(2): 185–190.

Nassar, N.N. 2013. The application of nanoparticles for wastewater remediation. Future Medicine.

Nel, A., Xia, T., Madler, L. and Li, N. 2006. Toxic potential of materials at the nanolevel. Science, 311(5761): 622–627.

Ortega-Liébana, M.C., Hueso, J.L., Ferdousi, S., Yeung, K.L. and Santamaria, J. 2016. Nitrogen-doped luminescent carbon nanodots for optimal photo-generation of hydroxyl radicals and visible-light expanded photo-catalysis. Diamond and Related Materials, 65: 176–182.

Peng, X., Li, Y., Luan, Z., Di, Z., Wang, H., Tian, B. et al. 2003. Adsorption of 1,2-dichlorobenzene from water to carbon nanotubes. Chemical Physics Letters, 376(1-2): 154–158.

Pervez, M., Balakrishnan, M., Hasan, S.W., Choo, K.-H., Zhao, Y., Cai, Y. et al. 2020. A critical review on nanomaterials membrane bioreactor (NMs-MBR) for wastewater treatment. NPJ Clean Water 3(1): 1–21.

Pidgeon, N., Porritt, J., Ryan, J., Seaton, A., Tendler, S., Welland, M. et al. 2004. Nanoscience and nanotechnologies: Opportunities and uncertainties. The Royal Society, The Royal Academy of Engineering 29(07): 2004.

Qu, X., Alvarez, P.J. and Li, Q. 2013. Applications of nanotechnology in water and wastewater treatment. Water Research 47(12): 3931–3946.

Rahmani, A., Ghaffari, H. and Samadi, M. 2011. A comparative study on arsenic (III) removal from aqueous solution using nano and micro sized zero-valent iron.

Rawal, S.B., Bera, S., Lee, D., Jang, D.-J. and Lee, W.I. 2013. Design of visible-light photocatalysts by coupling of narrow bandgap semiconductors and TiO 2: Effect of their relative energy band positions on the photocatalytic efficiency. Catalysis Science & Technology 3(7): 1822–1830.

Ren, X., Chen, C., Nagatsu, M. and Wang, X. 2011. Carbon nanotubes as adsorbents in environmental pollution management: A review. Chemical Engineering Journal, 170(2-3): 395–410.

Rivero-Huguet, M. and Marshall, W.D. 2009. Reduction of hexavalent chromium mediated by micron-and nano-scale zero-valent metallic particles. Journal of Environmental Monitoring 11(5): 1072–1079.

Roy, A. and Bhattacharya, J. 2015. Nanotechnology in Industrial Wastewater Treatment. IWA Publishing.

Salahi, A., Mohammadi, T., Behbahani, R.M. and Hemmati, M. 2015. Asymmetric polyethersulfone ultrafiltration membranes for oily wastewater treatment: Synthesis, characterization, ANFIS modeling, and performance. Journal of Environmental Chemical Engineering 3(1): 170–178.

Schmidt-Mende, L. and MacManus-Driscoll, J.L. 2007. ZnO–nanostructures, defects, and devices. Materials Today 10(5): 40–48.

Shirsath, D. and Shirivastava, V. 2015. Adsorptive removal of heavy metals by magnetic nanoadsorbent: An equilibrium and thermodynamic study. Applied Nanoscience, 5(8): 927–935.

Sreevidya, S., Subramanian, K.S., Katre, Y., Singh, A.K. and Singh, J. 2021. Functionalized Nanomaterial (FNM)–based catalytic materials for water resources. Functionalized Nanomaterials for Catalytic Application, 1–51.

Tristão, J.C., Magalhães, F., Corio, P. and Sansiviero, M.T.C. 2006. Electronic characterization and photocatalytic properties of CdS/TiO2 semiconductor composite. Journal of Photochemistry and Photobiology A: Chemistry 181(2-3): 152–157.

Venis, R.A. and Basu, O.D. 2021. Silver and zinc oxide nanoparticle disinfection in water treatment applications: Synergy and water quality influences. H2Open Journal 4(1): 114–128.

Wang, R., Guo, W., Li, X., Liu, Z., Liu, H. and Ding, S. 2017. Highly efficient MOF-based self-propelled micromotors for water purification. RSC Advances 7(67): 42462–42467.

Wenying, X., Ping, L. and Jinhong, F. 2008. Reduction of nitrobenzene by the catalyzed Fe/Cu process. Journal of Environmental Sciences 20(8): 915–921.

Zhao, X., Shi, Y., Wang, T., Cai, Y. and Jiang, G. 2008. Preparation of silica-magnetite nanoparticle mixed hemimicelle sorbents for extraction of several typical phenolic compounds from environmental water samples. Journal of Chromatography A 1188(2): 140–147.

Zodrow, K., Brunet, L., Mahendra, S., Li, D., Zhang, A., Li, Q. et al. 2009. Polysulfone ultrafiltration membranes impregnated with silver nanoparticles show improved biofouling resistance and virus removal. Water Research 43(3): 715–723.

CHAPTER 5

Water Purification and Role of Nanobiotechnology

Tariq Mehmood,[1,2] *Saira Bibi,*[3] *Muhammad Shafqat,*[3]
Beenish Mustafa,[4] *Licheng Peng,*[2,]* *Predrag Ilić,*[5]
Muhammad Anwar-ul-Haq,[6] *Mariym Sattar*[7]
and *Muhammad Faheem*[8]

1. Introduction

The greatest obstacle that must be overcome in the 21st century is ensuring that everyone has access to clean water to drink as the global population continues to rise (Gaurav et al. 2020, Mehmood et al. 2021a). It has been estimated that 10–20 million people per year succumb to death due to diseases transmitted through water (Leonard et al. 2003). According to the estimation, almost one billion people worldwide will not have safe

---

[1] Helmholtz Centre for Environmental Research - UFZ, Department of Environmental Engineering, Permoserstr. 15, D-04318 Leipzig, Germany.
[2] College of Ecology and Environment, Hainan University, Haikou, Hainan Province, P.R. China 570228.
[3] Pak-Austria Fachhochschule, Institute of Applied Science and Technology, Mang, Haripur, Khyber Pakhtunkhwa, 24421, Pakistan.
[4] National Laboratory of Solid State Microstructures, School of Physics, Nanjing University, Nanjing 210093, China.
[5] PSRI Institute for Protection and Ecology of the Republic of Srpska, Vidovdanska 43, Banja Luka, Republic of Srpska, Bosnia and Herzegovina.
[6] Department of Horticulture, Bahauddin Zakariya University Multan, Pakistan.
[7] Department of Aeronautics and Astronautics, Institute of Space Technology, Islamabad, 2750, Pakistan.
[8] Department of Civil Infrastructure and Environment Engineering, Khalifa University of Science and Technology, 127788 Abu Dhabi, United Arab Emirates.
* Corresponding author: lcpeng@hainanu.edu.cn

and clean drinking water in a few decades, when water supply will be reduced by one-third from the current availability, due to the lack of water reserves. When water supply is likely to be reduced, this will happen. The rapid increase in the world's population, coupled with the unplanned urbanization and industrialization that has accompanied it, is a major contributor to the pollution of natural resources such as air, water and soil (Mehmood et al. 2021b, Mehmood et al. 2020). The massive amounts of wastewater produced by households and businesses contribute to contaminating water supplies with dangerous pollutants such as radioactive materials, metalloids, nitrates and heavy metals (Mustafa et al. 2022). The quality of the nearby water supply is significantly impacted by the usage of pesticides and other farming chemicals to increase production. These hazardous chemicals and heavy elements in water reserves have a lethal effect on not only humans but also aquatic animals and other forms of life. Therefore, if we have to keep up with the expanding demand, we obviously need to create more sophisticated systems for drawing drinkable water from rivers and reservoirs. Reverse osmosis, adsorption, electro-floatation, ion exchange, nanotechnology and membrane separation (Sharma et al. 2019) are examples of effective effluent treatment techniques that must be used if these process sectors are to survive. Among these methods, the use of nanotechnology-based water treatment techniques is seeing tremendous growth as these methods are both very effective and viable (Qu et al. 2013a, Qu et al. 2013b).

Nanomaterials are materials with particles typically no larger than one billionth of a meter. A comparison of nanomaterials with other large-sized particles is depicted in Figure 1.

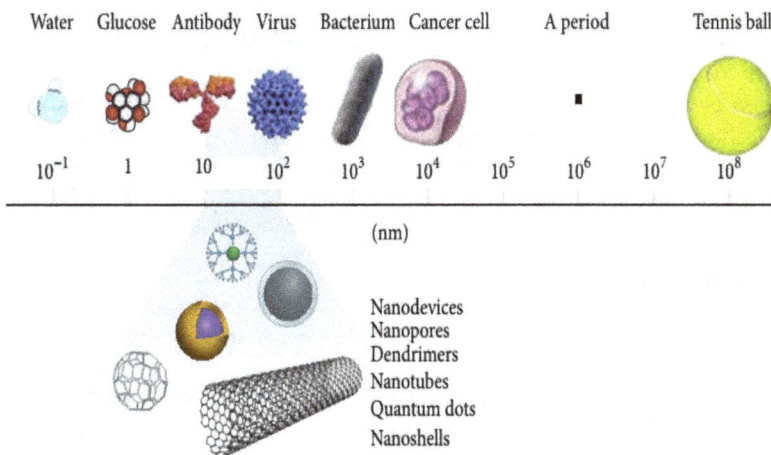

Figure 1: Comparison of nanomaterials with other large-sized particles. Reproduced from Amin et al. (2014).

In most cases, the physical, chemical and biological properties these materials are used to define their characteristics. Nanoparticles and macroparticles differ fundamentally in terms of their properties–magnetic, electrical and optical. The exceptionally high number of atoms on the surface of the nanoparticles has given rise to their one-of-a-kind properties. In addition, nanomaterials have size-dependent properties that are unique and distinct from those of their large particle counterparts; they are hence being investigated for potential use in the purification of polluted water (Karri et al. 2018, Lingamdinne et al. 2018). NBT is a technique that mainly utilizes the advantages of nanotechnology as well as the properties of biomolecules found in living cells. Nanotechnology has the potential to be used to develop new goods by applying its principles and traits to a variety of biological processes. The procedures presented here are well-illustrated by techniques like membranes and filters, nano biosensors, nanofluids and nano proteomics. It has been hypothesized that NBT might hold the key to solving the problems we currently face with water, food and biodiversity.

These days, research efforts have been centered around finding ways to use NBT to treat water, instead of relying on the conventional methods that are currently in use. It has been hypothesized that the application of nanotechnology to the treatment of water can result in a process that is both effective and versatile and does not rely on large infrastructures (Qu et al. 2013b). Other methods of water purification, such as coagulation (Sillanpää et al. 2018), membrane technology (Ang et al. 2015), adsorption and photocatalysis (Dong et al. 2015), are all compatible with this technology and can be used in conjunction with it. Additionally, the scientific discipline of nanotechnology, when combined with engineering technologies, has been supposed to be extraordinarily helpful. These techniques are used to address problems associated with the quality of water through nanostructured catalytic membranes, utilizing nanotubes, nano-adsorbents, nano-catalysts and macromolecules. (Gupta et al. 2006). This chapter summarizes the use of materials based on NBT for the treatment of industrial wastewater, either on their own or in combination with other technologies.

2. Nanobiotechnology

The application of nanotechnology to the study of biological systems is known as nanobiotechnology (NBT). Researchers in chemistry, physics as well as biology consider nanotechnology to be a subfield of their respective fields; hence, it is not uncommon for these three groups to work together on projects to which they each contribute equally. One of these consequences is the emergence of a hybrid field known as

NBT, which utilizes biological building blocks and designs ideas or has applications in the biological or medical sciences. NBT, a combination of biotechnology and nanotechnology, is an emerging field that focuses on other physiological processes of living things including metabolic process of microorganisms. This field is expected to play an important role in developing and applying an extensive array of helpful tools for studying life.

Research in the field of nanobiotechnology is still in its infancy, although the combination of biology and nanomaterials has already resulted in the creation of diagnostic devices, drug-delivery vehicles, analytical tools, contrast agents and contrast agents therapy.

3. Biogenic Nanoparticles

Biogenic nanoparticles are created by living things like yeast, bacteria, fungi, lilies, algae and plants. Because of their nontoxicity, biologically produced nanoparticles have become a viable proxy to chemically produced nanoparticles. Recently, several nanoparticles derived from biological sources have been developed and could be used for environmental remediation and healthcare purposes. Numerous studies have shown that biogenic nanoparticles, such as biogenic nano-magnets, bio-palladium nanocrystals, biogenic manganese oxide (BioMnOx) and biogenic iron species, are effective in removing micropollutants, heavy metals, halogenated chemicals and refractory pollutants. Environmental remediation could thus benefit greatly from nano-bioremediation – a new, safer, more environmentally friendly and more cost-effective technology. The uses of microbial nanoparticles in synthesis, organization and environmental bioremediation have been reviewed in this chapter.

Nanoparticles fabricated through the use of fungi, yeast, algae and bacteria nanoparticles that plants produce are known as biogenic nanoparticles (Yadav et al. 2017). The creation of nanoparticles by biological processes is known as 'bio-nano factories', and this process is the main driver of motivation. These living things make proteins in their cytoplasm and are composed of biomolecules.

This then results in the depletion of metallic ions, ultimately synthesizing nanoparticles (Rai et al. 2011). These biogenic nanoparticles have potential to use in a wide variety of fields, including chemical and electrochemical sciences, the medical field, food industry, biotechnology and environmental remediation (Sharma et al. 2019). It has been discovered that these nanoparticles have a wide range of applications involving the treatment of wastewater, remediation of soil and air, disinfection of water, as well as sensing and detection, seeing that groundwater and surface water pollution have increased dramatically due to extensive urbanization and industrialization.

About 2.4 million metric tons of waste containing heavy metals is produced annually by industries related to fuel and power production; agriculture and waste disposal contribute an additional 2 million metric tons annually (Dhillon et al. 2012). Typically, organic and inorganic pollutants make their way into the environment, where they undergo transformations that result in the formation of toxic compounds and contaminate water bodies and soil. The Environmental Protection Agency (EPA) has identified the industrial manufacturing sector as the primary route through which heavy metals such as lead (Pb), arsenic (As), mercury (Hg) and cadmium (Cd) are released into the environment. These heavy metals are considered to be critical pollutants. In addition, the high intrinsic toxicity of several other pollutants like trichloroethylene (TCE) and perchloroethylene (PCE) makes biological remediation of these pollutants extremely difficult. Another substance to consider here is polynuclear aromatic hydrocarbons (PAHs)–they present a similar problem (Gaurav et al. 2021). It has been demonstrated that nanoparticles are adequate for the removal of a variety of pollutants, including dyes (Aziz et al. 2015), pesticides (Das et al. 2009), halogenated compounds (Hennebel et al. 2009), pharmaceutical products and heavy metals (Castro et al. 2018) from the soil and water matrix.

Environmentally friendly biogenic nanoparticles have shown to be a successful and evolving solution for the removal of environmental toxins, given their excellent catalytic and adsorptive capabilities. This chapter presents a comprehensive discussion on the synthesis of nanoparticles through the use of microorganisms, and the function these organisms play in environmental remediation (Figure 2). Regarding removing toxins from water and soil, several biogenic nanoparticles and synthetic nanoparticles employing just microorganisms have also been addressed for their function and mechanisms of action. The limitations associated with biogenic nanoparticles are also discussed in this study. These limitations could be circumvented for a more effective application to eliminate various environmental contaminants.

4. Benefits Associated with Using Biogenic Nanoparticles

The natural world has provided alternative routes that are clean and environment friendly for the synthesis of nanoscale materials. Nanoparticles produced through biogenic agents such as algae, bacteria and fungi are, in many respects, a better and more successful replacement to those produced through chemical and physical methods (Figure 2). Because they use biomolecules as reducing agents, biogenic

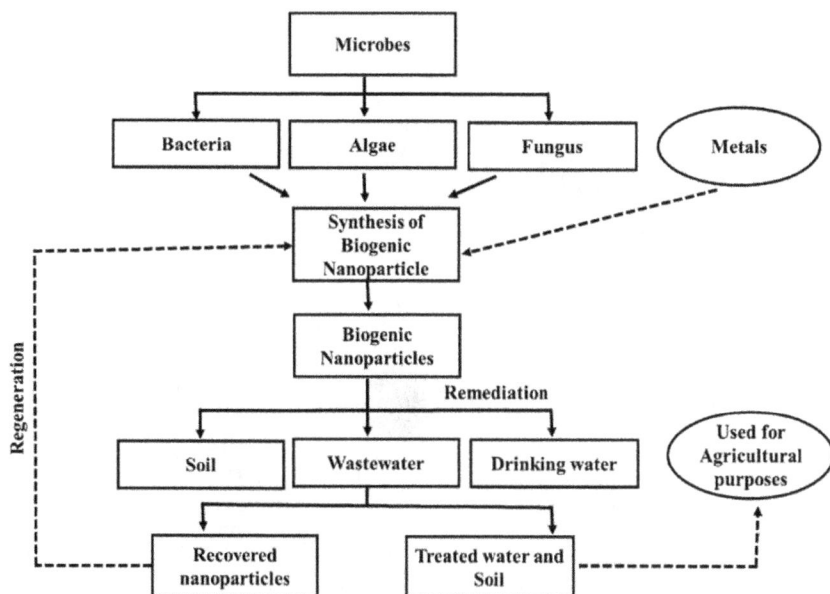

Figure 2: Synthesis of biogenic NPs and their uses. Reproduced from Kumari et al. (2019) with permission. Copyrights Springer Nature 2023.

nanoparticles are less expensive than nanoparticles produced by applying physicochemical methods. This is because biomolecules eliminate the need of costly chemical reductants like sodium borohydride and hydrazine. The production of nanoparticles through chemical processes generates toxic wastes which are detrimental to the well-being of mankind and the environment. In contrast, the generation of nanoparticles through biogenic mechanisms does not result in the generation of such hazardous toxic wastes. Furthermore, nanoparticles produced using biogenic routes have a larger surface area; this significantly improves their adsorption capacity for the elimination of atmospheric pollutants.

Most biogenic nanoparticles have lipid bilayers which impart to them the properties of better physiological solubility and stability. Figure 3 shows various properties of biogenic nanoparticles. These qualities make them suitable for various applications in the field of biomedicine. In addition, the pH, availability of substrates and the amount of time that the reactants are in contact with one another can be altered to influence the shape and size of the nanoparticles (Li et al. 2011). Therefore, biogenic nanoparticles have emerged as cost-efficient, eco-friendly and superior alternatives to physicochemical methods, and have a less energy-intensive synthesis as well. This is because biogenic nanoparticles are superior in terms of cost and their impact on human health and the environment.

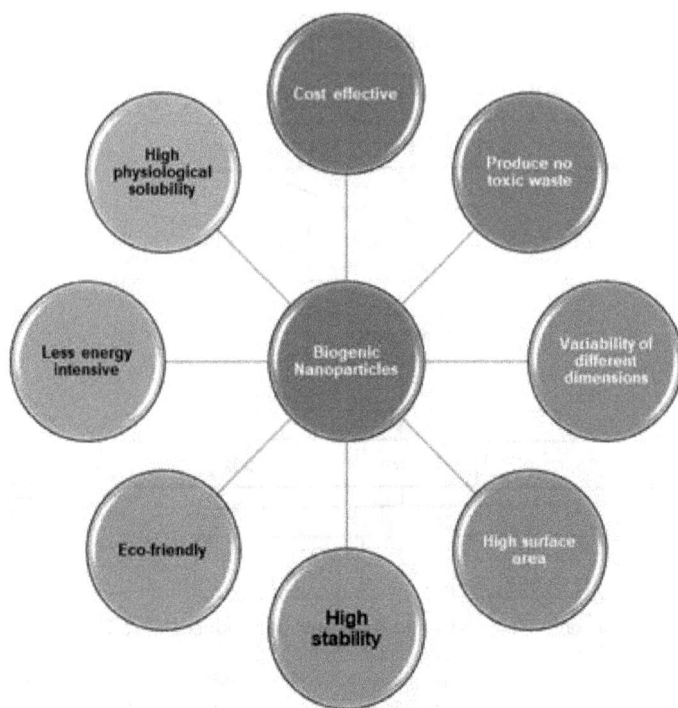

Figure 3: Properties of biogenic NPs.

5. Biological Synthesis of Nanoparticles

Nanoparticles can also be produced by microbes using either extracellular or intracellular processes to complete the synthesis. Electrostatic interactions allow for the diffusion of cations into the anionic cell wall during the processes that take place inside the cell. Carrier channels, endocytosis and ion channels are all mechanisms that contribute to the facilitation of the entry of heavy metals into the cytoplasm. Afterwards, these harmful metals are converted into non-toxic nanoparticles by the enzymatic machinery in the cell wall. Enzymes like nitrate reductase, excreted by fungi and which convert metal ions into metal nanoparticles, are examples of extracellular activities. This process is an example of an extracellular catalysis (Menon et al. 2017). Characterization of nanoparticles through various approaches is typically performed in conjunction with the biosynthesis of nanoparticles. An accurate picture of the size, shape, chemical composition, surface area and dispersity of nanoparticles can be obtained through the characterization of these particles.

Scanning electron microscopy (SEM), X-ray diffraction (XRD), energy-dispersive X-ray spectroscopy (EDS), X-ray photoelectron spectroscopy (XPS), transmission electron microscopy (TEM), atomic force microscopy (AFM), ultraviolet-visible spectroscopic analysis and Fourier-transform infrared spectroscopy (FTIR) are frequently used to characterize synthesized nanoparticles.

5.1 Bacterial Synthesis of Nanoparticles

Different types of bacteria which have been demonstrated to have the capacity to operate as biofactories can produce nanoparticles such as silver (Ag), gold (Au), iron (Fe), copper (Cu), zinc (Zn), palladium (Pd), manganese (Mn), and several other designed nanoparticles (Table 1). Bacteria can organize, disable and precipitate metals, which makes it simpler to synthesize nanoparticles when they are present. When put down in a silver nitrate solution, *Pseudomonas stutzeri* AG259, which was secluded from silver mines and used in a study reported by Klaus et al. was able to lower the concentration of silver ions (Klaus et al. 1999). The accumulated silver nanoparticle could then be seen within the periplasmic space of the cell after this process had taken place. *Rhodopseudomonas capsulata* has been found to be responsible for forming spherical gold nanoparticles ranging in size from 10 nm to 20 nm, when grown in an acidic environment (He et al. 2007). According to Sundaram et al. (2012), the bacterium *Bacillus subtilis*, obtained from soil samples taken from rhizospheres, possesses the ability to biosynthesize iron oxide nanoparticles.

Table 1: Synthesis of biogenic nanoparticles by employing bacteria. Adapted from Kumari et al. (2019).

Type of NPs	Bacteria
Gold	*Streptomyces* sp. *VITDDK3, Rhodopseudomonas capsulate, Pseudomonas aeruginosa, Thermomonospora* sp., *Klebsiella pneumonia* and *Brevibacterium casei*
Iron	*Bacillus subtilis* and *Klebsiella oxytoca*
Copper	*Psedumonas stutzeri* and *Pseudomonas fluorescence*
Silver	*Staphylococcus aureus, Escherichia coli, Stryptomyces rochei* and *Bacillus thuringiensis*
Zinc	*Lactobacillus plantarum VITES07* and *Actinomycetes*
Biogenic manganese oxide NPs	*Pseudomonas putida*

5.2 Algae-based Synthesis of Nanoparticles

Microorganisms that belong to the kingdom Eukaryote, are photoautotrophic and oxygenic, and possess the ability to accumulate metals are called algae. The most popular algae species used for the synthesis of nanoparticles are listed in Table 2. Many advantages are associated with using algae for the bio-preparation of metallic nanoparticles, including the capacity to bioaccumulate metals, a high tolerance, ease of handling, and the financial sustainability of the procedure. Moreover, the production of enzymes on an industrial scale is made possible by algae's extensive secretion of extracellular enzymes (Thakkar et al. 2010). Fucoidans are polysaccharides that are secreted by marine brown algae from their cell walls. It has been seen that fucoidans have anti-inflammatory, antiviral, anticancer and anticoagulant effects. Fucoidans are said to be a key component in the creation of gold nanoparticles, offering a more ecologically friendly alternative to conventional processes that primarily rely on chemicals (Lirdprapamongkol et al. 2010). Brown algae have abundant mucilaginous polysaccharides and carboxyl groups, both of which are essential for metal uptake. As a result of their capacity to absorb metals, brown algal cell walls are effectively commonly used in the production of nanoparticles (Khandel and Shahi 2016).

Table 2: Synthesis of biogenic nanoparticles by employing algae. Adapted from Kumari et al. (2019).

Type of NPs	Algae
Gold	*Tetraselmis kochinensis, Chlorella vulgaris, Lemanea fluviatilis, Turbinaria conoides, Sargasum muticum* and *Padina gymnospora*
Iron	*Sargassum muticum* and *Euglena gracilus*
Copper	*Bifurcaria bifurcate*
Silver	*Turbinaria conoides* and *Chlamydomonas reinhardtti*
Zinc	*Sargassum muticum*

5.3 Nanoparticle Synthesis using Fungi and Yeast

In the production of nanoparticles, fungi have also been utilized to a significant degree (Table 2). They are simple to use and a reliable source of extracellular enzymes which ease the creation of nanoparticles. When compared to bacteria and algae, fungi are more economically viable for use in the synthesis of nanoparticles and produce significant amounts of nanoparticles. Furthermore, filamentous fungus is easy to process in downstream stages, has a potential economic value, and has a high capacity for bioaccumulation and high tolerance for metals. It has been demonstrated that fungi can produce gold nanoparticles inside their cells. In one experiment, gold nanoparticles were located inside the cell vacuoles of fungus cells in ultra-thin sections (Menon et al. 2017). Also, verticillium has been found to have an intracellular preparation of silver nanoparticles (AgNPs) (Mukherjee et al. 2001). By exposing fungal biomass to an aqueous solution carrying Ag⁺, metal ions were reduced and AgNPs with dimensions of 25 nm × 12 nm were produced. The morphology of the artificially produced silver nanoparticles was further established to be spherical by transmission electron microscopy (TEM) images.

Only a small number of yeast strains have been employed to create nanoparticles till now. In one study, two different strains of the yeast Saccharomyces cerevisiae were used to produce Au (gold) nanoparticles (Sen et al. 2012). Low dosages of γ-energy was applied to the cell's cytoplasm to facilitate the in-situ reduction of Au^{3+} to Au (0), and nanosized Au^{3+} was discovered inside the nucleolus. The production of gold nanoparticles has been accomplished by employing yeast-derived reductants as a component of the reagent.

6. Applications of Biogenic Nanoparticles

Not long ago, there was a rise in the interest in environmental remediation based on NBT. As a biological entity, a microorganism is considered as a viable source because it is simple to work with, inexpensive and tolerant of high levels of metals (Narayanan et al. 2013). A sustainable method for the elimination of environmental pollutants involves utilizing a microbial platform in the production of metal nanoparticles. To address a wide range of environmental concerns, several research projects taking place in laboratories as well as applications of biogenic nanoparticles in the field are currently underway. Recent developments in nano-bioremediation involve removing heavy elements and organic and inorganic pollutants from surface water, groundwater, soil and wastewater; the said contaminants can be found in all four of these environments. The biological method of producing nanoparticles results in particles with high levels of reactivity and a surface area that is significantly larger than that of chemically produced nanoparticles. These biologically produced nanoparticles have been shown to act as reductants or oxidants, adsorbents and catalysts. The production of nanoparticles frequently involves the use of bacteria capable of either reducing or oxidizing iron.

Biogenic Fe oxides not only separate metals like As, chromium and cobalt through a process known as adsorption, but also cut down on organic contaminants known as nitro-aromatics and encourage the dichlorination of aliphatic compounds. Owing to oxidation and reduction reactions, iron oxides are an important component in the removal of contaminants (Castro et al. 2018). The biological synthesis of silver nanoparticles has been the subject of significant research, and these nanoparticles have been reported to possess beneficial antibacterial and catalytic properties.

Photocatalytic degradation of organic pollutants using silver nanoparticles has emerged as a method that is both cost-effective as well as kind to the environment for the abolition of poisonous organic contaminants from the environment (Aziz et al. 2015).

6.1 Removal of Heavy Metals from Wastewater

Because of their teratogenic and carcinogenic characteristics, heavy metals are bad for both humans and other living things. These characteristics have detrimental effects on human health. It is, therefore, of utmost importance to locate a path that removes heavy metals from the environment in an environmentally friendly manner. For example, procedures connected to mining and the processing of metallurgical materials commonly produce wastewater that contains heavy metals like As and Cu. According to a recent study (Castro et al. 2018), As, Cu, Zn and Cr may all be removed from wastewater using iron nanoparticles made from natural consortia. Magnetite (Fe_3O_4), vivianite ($Fe_3(PO_4)_2 8H_2O$) and siderite were the three types of biological precipitates that were obtained from the $FeCO_3$ cultures. These biogenic adsorbents demonstrated a higher ability to adsorb arsenate (AsV) and a higher affinity for AsV than chromate. The organic components affixed to the nanoparticle adsorbed both Cu and Zn.

High levels of As, iron (II) and manganese (II) in groundwater and surface waters harm aquatic ecosystems and human health. One strain of Pseudomonas sp. QJX-1 synthesized biogenic Fe–Mn oxides by oxidizing manganese (II) (BFMO). As was found to be oxidized and adsorbable by these oxides. First, the BFMO is oxidized chemically and biologically to FeOOH and MnO_2 on the surface of BFMO to form Fe–Mn oxide which adsorbs and oxidizes the As FeOOH has a higher capacity for adsorption but a lower capacity for oxidation, which is counterbalanced by higher oxidation efficiency of biological manganese oxide (BMO). FeOOH and BMO create BFMO; this promotes the appropriate adsorption and oxidation of As (III) and As (V). BMO can simultaneously lower the pollutant levels of As (III or V), Fe (II) and Mn (II) by oxidizing Fe (II) to FeOOH which subsequently adsorbs any leftover As (III) or As (V) in the groundwater (Figure 4).

Using SRB to treat chromium-containing wastewater is another option. Sulfate (SO_4^{2-}) concentration and chemical oxygen demand (COD) of wastewater can be reduced by using organic compounds as a carbon source for these bacteria. Under ideal circumstances, sorption trials on simulated wastewater reduced COD, Cr (VI) and sulfate from synthetic wastewater by 81.9 percent, 89.2 percent and 95.3 percent respectively. SRB has also been used in the bioremediation of As-contaminated groundwater (Saunders et al. 2018). Evidence suggests that biogenic Fesulfides can remove Zn ions from mine drainage water, meaning that this technique could also be used to clean up genuine sewage. SRB works

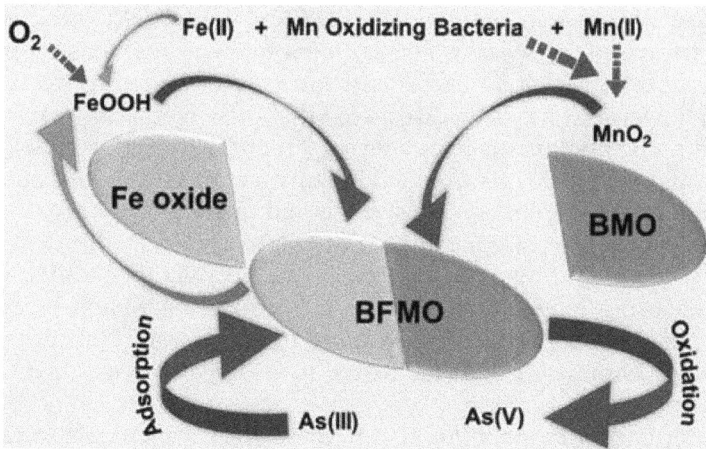

Figure 4: Proposed mechanism of adsorption and oxidation of As (III) by BFMO.

well as a bio-adsorbent to remove sulphate and heavy metals/elements from groundwater and wastewater.

Heavy metallic elements can be removed from the atmosphere through biogenic manganese oxide, also known as BMO, produced from *Pseudomonas putida* MnB1. BMO is superior to chemically synthesized manganese oxide as a heavy metal adsorbent (Bri). Such BMO is different from chemically manufactured Bri; whereas the latter is crystalline and has a stick-like morphology, the former is amorphous, uneven and nanosized. BMO is a powerful adsorbent because of its amorphous shape, tiny size and high surface area. It has seven to eight times the adsorption capacity for Zn, Pb and Cd, compared to birnessite (Bir). Additionally, when the pH and temperature are raised from three to six and from fifteen to thirty degrees Celsius, respectively, the adsorption capabilities of heavy metals rise. Because adsorption capacity increases with temperature, it is also possible that the adsorption process is endothermic at its core.

6.2 Removal of Organic Pollutants

The elimination of organic micropollutants using biogenic nanoparticles is the subject of a great deal of research these days. This includes the removal of halogenated impurities, organophosphorus pesticides, dyes and pharmaceutical products.

6.2.1 Removal of Pharmaceutical Products

There is a severe issue involving the discharge of pharmaceutical goods and other organic substances into oceanic habitats that has drawn the attention of researchers worldwide (Kümmerer 2009). The original goal– treatment of wastewater by conventional methods– was the reduction of COD, BOD and other nutrients in larger concentrations (ppm); however, it has been discovered that these methods are unable to deal with micropollutants in ppb levels. According to Ternes et al. (2004), wastewater, and effluents from sludge treatment facilities are sources of contaminants that enter aquatic systems and their surroundings in lower concentrations. The steroid hormones estrone and 17-ethinylestradiol, diclofenac and ibuprofen, among other organic micropollutants, can be removed using biogenic manganese (Mn) oxide nanoparticles created using *Pseudomonas putida*. Estrone and 17-ethinylestradiol are almost completely eliminated with the use of biogenic manganese (Mn) oxide nanoparticles.

According to Furgal et al. (2015), these particles are able to remove organic micropollutants even when they were in lower concentrations (at the ppb level). Both BioMnOx and Bio-Pd have biologically generated nanoparticles that can remove several micropollutants from waste treatment plant effluents. In the said study, BioMnOx was synthesized by the *Pseudomonas putida* MnB6 strain, whereas bio palladium (Bio-Pd) was synthesized by the bacterium *Shewan Ella oneidensis*. These nanoparticles were used for the oxidation–reduction of pollutants in laboratory-scale membrane bioreactors (MBR).

6.2.2 Removal of Dyes and Other Contaminates from Wastewater

It has been discovered that the palladium (0) nanoparticle, also known as bio-Pd, can reduce the azo dye, when synthesized using *Klebsiella oxytoca* GS-4-08 under fermentative conditions. This may be useful in the process of designing and operating bioreactors that remove azo dye from wastewater (Wang et al. 2017). Methylene blue, also known as MB, is a tenacious organic compound that is extremely hazardous to the health of aquatic organisms. Due to the high toxicity of methylene blue, its removal through biological processes is extremely difficult.

Aziz et al. (2015) described the use of an algal platform consisting of *Chlorella pyrenoidosa* to manufacture biogenic silver nanoparticles. They tested the photo-catalytic efficiency of biologically produced silver nanoparticles (biogenic AgNPs) in the MB dye degrading process. Compared to silver nanoparticles produced through commercial synthesis methods, the photocatalytic degradation of methylene blue by biogenic AgNPs was significantly superior. Irradiation of AgNPs with visible light is what kickstarts the photocatalytic activity of these nanoparticles, making it possible for photoelectrons to move from the valance band of the nanoparticles to the conduction band. A hole pair (h^+) in the valance band and electron-hole (e) in the conduction band are produced as a result, along with a band gap. When a high band gap is present, the development of non-radiative recombination of electron holes and hole pairs, which are effective oxidizing and reducing agents, increases the photocatalytic activity of silver nanoparticles. In the direction of the conduction band, oxygen binds to nanoparticles, where it can capture electrons and reduce them to form the anionic superoxide radical (O_2). When hydrogen peroxide (H_2O_2) is created due to the newly formed superoxide radical's ability to take protons, it again splits into hydroxyl radicals (OH). Water is simultaneously oxidized in the valence band on the other side by hole pairs and is sucked up by them, creating hydroxyl radicals. All of this occurs simultaneously. Anionic superoxide and hydroxyl radicals attack the aromatic rings, azo linkages and hydroxylated rings, causing MB to break into CO_2, H_2O, SO_4^{2-}, NH_4^+ ions and NO_3^-.

Nitroaromatic compounds are examples of highly poisonous impurities mostly found in the effluents from dye manufacturing facilities and eventually make their way into the bodies of water in the surrounding area. Because of its high solubility and stability in water, 4-nitrophenol poses a significant threat to the environment and public health because of its cancer-causing and mutation-causing properties. They take a long time to decompose, and some may stay buried in the ground for an unforeseen period. Nitrophenols are considered 'priority pollutants' by the United States Environmental Protection Agency (USEPA); the latter has placed a maximum permissible concentration of about 20 parts per billion for them. It is thus required to efficiently decompose these contaminants.

According to Narayanan and Sakthivel (2011), biogenic gold nanoparticles produced from *Cylindrocladium floridanum* and filamentous fungal biomass are useful in the catalytic dissociation of toxic 4-nitrophenols into non-toxic 4-aminophenols.

6.2.3 Disinfection and Removal of Organophosphorus Pesticides from Water

In a one-step treatment process, nanogold bioconjugate (NGBC) manufactured using the *Rhizopus oryzae* strain has shown to be effective in the removal of organophosphorus pesticides as well as certain microorganisms from water (Das et al. 2009).

In the abovementioned study, experiments in the laboratory were carried out using simulated wastewater that contained *E. coli* at a concentration of at least 103 cells per milliliter, in addition to malathion (10 g/L), parathion (5 g/L), chlorpyrifos (12 g/L) and dimethoate (8 g/L). After adding five grams of NGBC to one hundred milliliters of this wastewater, the mixture was heated to thirty degrees Celsius and incubated for varying amounts of time. After the allotted time had elapsed in the incubation process, the NGBC was filtered sterilely, using glass wool. Following filtration, the cell count of *E. coli* and the concentration of the pesticide in the filtrate were determined. Within 10 minutes, it was discovered that the concentration of the pesticides and the density of *E. coli* in the treated water decreased significantly compared to the control water. After thirty minutes of incubation, the pesticide level was determined to be way below the detection limit (less than one microgram per liter), and all traces of *E. coli* were eliminated. The *E. coli* was not eradicated in the control experiment that used *pure R. oryzae mycelia* (which lacked an embedded gold nanoparticle). In elemental compounds mercury selenium compound (HgSe) is more common. This HgSe is less toxic and chemically inert than its reactive counterpart. This has led to the conclusion that NGBC can be used. In addition, nano-selenium (0) in the soil allows for capturing Hg (0) in both aerobic or anaerobic environments. As a result, making use of nano-selenium for the immobilization of mercury is an effective method for cleaning up soils that are contaminated with mercury.

6.3 Bioactive Nanoparticles

Bioactive materials are a type of material with distinctive bioactivities. These materials can alter the behaviors of living cells and elicit biological responses from living tissues. In the late 1960s, bioactive materials began to receive significant attention. It is a common practice to use silver nanoparticles in water purification systems to eliminate bacteria and other microbe-based contaminants. Scientists have discovered that nanoparticles can also be produced from biological sources. This was an unexpected finding. Silver nanoparticles (AgNPs) produced by synthesizing the extracellular material of the bacterium *Bacillus cereus* have been shown to have very high antibacterial efficacy against both Gram-negative and Gram-positive bacteria. According to Choi et al. (2014), magnesium (Mg) oxide nanoparticles and cellulose acetate (CA) fibers with embedded silver (Ag) nanoparticles are two of the most successful biocides against Gram-negative and Gram-positive bacteria respectively.

6.4 Use of Biomembranes for the Treatment of Water

Bio-membrane-based water H_2O treatment is one of the most recent and advanced techniques for water purification due to the distinctive ways in which it is built and designed. University of New Mexico as well as Albuquerque's Sandia National Laboratories have made substantial contributions to the creation of this biomimetic membrane. They claim that the self-assembling nanopores used in the invention were optimized using atomic layer deposition and that these nanopores generally provide high flux desalination. By applying pressure and utilizing the appropriate amount of electrical energy, the membranes are able to filter impurities like salt, etc. from the water. The nonporous biomimetic design of the membrane allows for high salt rejection at low-pressure levels (5.5 bar), and water flows faster through the membrane. Basically, the principle of reverse osmosis was followed in the filtration process being discussed. Due to the fact that this particular method combines nanofabrication with the protein channels found in biomembranes, it can achieve a high level of filtration efficiency while maintaining a low operational expenditure (Choi et al. 2014).

7. Biomaterials and Polymers

In addition to higher sensitivity and selectivity, biosensors based on polymers and biomaterials also provide a quick response time due to their electrical, radiant, mechanical, catalytic and thermal capabilities (Yang et al. 2016). By combining scientific methods and novel apparatus such as high-throughput material screening with microfluidics and micro/ nanofabrication, it is now possible to plan and develop a biosensor for determining environmental impurities in water, based on polymeric biomaterials with special properties and functionalities. Polymeric biomaterials with distinct functionalities and characteristics would serve as the foundation for the said biosensor (Yuan et al. 2021).

7.1 Polymeric Materials

The progress of polymeric materials, such as formulated materials, homo- and copolymers, polymeric structures with modified morphology, and molecular shape recognition materials, has been settled for the careful biosensing of environmental impurities in water (Chamjangali et al. 2015). Polymeric materials have demonstrated a range of qualities in various applications, including sensitivity, linearity, ease of manufacture, and selectivity. Developing self-assembling biomembranes, dendrimer-based polymeric materials and electropolymerized polymers has greatly influenced biosensor design. Polyfunctional groups can be seen on the surface of organic human-made polymers or biopolymers like chitosan and cellulose. These groupings include, among others, $^{-}NH_2$, ^{-}OH, and $^{-}COOH$ (Tehrani et al. 2013). However, their applicability is constrained due to their modest surface area and low adsorption rate. Meanwhile, materials like carbon or metal nanoparticles have a high surface area and can be chemically functionalized. This enhances biocompatibility, sensitivity and selectivity (Xue et al. 2013).

7.2 Biomaterials

The incorporation of biomolecules into nanocomposites brings together several advantageous properties that are unique to both nanomaterials and biomolecules; coupling the functional groups in nanocomposites with biomolecules such as enzymes or antiseptics is possible. Because of their distinctive recognition and electronic, transport and catalytic capabilities, biomaterials provide high recognition and selective catalytic properties (Sabela et al. 2016). Polymeric nanocomposites have larger surface areas and contain more functional groups on their surfaces than nanoparticles consisting of metals or oxides, organic polymers and activated carbon. The terms 'polymer nanoparticles', 'polymer nanofibers', 'polymer nanocrystals' and 'polymer nanorods' all refer to different types of polymer biomaterials. The creation and use of polymer biomaterials offer a new possibility for accurately detecting impurities in water. This has been made possible by recent advances in the field (Li et al. 2015).

7.3 Bionanocomposites

Bionanocomposites are a type of natural nanocomposite. Adaptability, modularity and multifunctionality are all characteristics of natural bionanocomposites. To ensure the long-term survival of the species with which they are associated, they are created specifically considering the needs of life and the surrounding environment. Biologists have studied their structures and properties extensively to understand the functions of biological polymers better. As a result, when designing binanocomposites, it is necessary to think of biological molecules as synthetic 'building blocks' that may be used in a context that is different from their natural setting or function. The structure and composition of raw materials are constantly changing. Ontogenesis and morphogenesis, two fundamental biological processes, are connected to these basic principles. It is possible that both these processes can lead to new materials with unforeseen properties and new applications.

8. Applications of Bionanocomposites in Wastewater

Herbicides, pesticides and insecticides are all examples of organic compounds essential for degrading environmentally hazardous contaminants; however, an excessive amount of these compounds have made their way into drinking water. When it comes to eliminating contaminants of this kind, organically modified CNC, particularly bio-nano clay composites, are the most environmentally friendly nanomaterials.

Alterations in the pH of the solution affects the adsorption capacity of these bionanocomposites. By bringing the pH closer to the acidic side, the transfer percentage of clopyralid—a herbicide produced by MMT-Chitosan NC—increases by 5–10 percent (Celis et al. 2012). According to the adsorption isotherms, the utilization of MMT surfactant reformed with trimethyloctylammonium bromide (TMOAB) results in a very high adsorption capacity for diazinon. This capacity has been reported to be 1428.5 mg g^{-1} when the MMT surfactant is used. The results of experiments have shown that the maximum removal efficiency can be accomplished in just sixty minutes using 0.5 grams of adsorbent per liter (Hassani et al. 2015). Under the influence of ultraviolet light, TiO$_2$ NP-doped MMT and hectorite synthesized using the sol-gel method catalytically degrade dimethachlor. The amount of doped TiO$_2$ determines the rate of degradation, which in turn affects the particle size of the NP in CNC, the pore volume and the surface area. A sequence is observed in adsorption: hectorite with 15% TiO$_2$ > hectorite with thirty percent TiO$_2$ > hectorite with 55% titanium dioxide > MMT with 15 percent TiO$_2$ > MMT with 30 percent TiO$_2$ > MMT with 55 percent TiO$_2$ > TiO$_2$; this indicates that TiO$_2$-NC of hectorite manifests a higher activity, compared to TiO$_2$-MMT; the reason for this behavior has been associated with two factors: (1) difficulty in gaining access to pesticides within the MMT pillared structure, and (2) efficient charge transfer within the hectorite NC. In general, all prepared NCs showed a very low percentage of adsorption – less than twenty percent (Belessi et al. 2007).

9. Integration of NBT and Existing Technologies

9.1 Nanoparticle-based Algae Membrane Bioreactor

Growing micro- and macroalgae and in wastewater to produce energy based on biomass and water purification has been acknowledged as a potentially successful tactic. Due to the abundance of the necessary micro- and macronutrients, many algae species can grow successfully in wastewater. This procedure concurrently eliminates nutrients from the wastewater since algal development requires nutrients, light and CO$_2$. Various processes involving centrifugation with sedimentation, chemical flocculation and air flotation can be used to harvest the biomass after the algal growth reaches a threshold, although these are not practical for usage on a large scale.

The membrane technology which enables the capacity to culture and reaps high-density algae in a simple membrane bioreactor is the most hopeful of techniques that have been created for the farming and harvesting of algae (Mehmood et al. 2022a, Mehmood et al. 2022b). Other methods for growing and gathering algae include the membrane technology which does not call for the inclusion of chemicals like coagulants during the membrane filtering process, unlike other harvesting techniques, which makes it simpler to reuse water after 'filtration and biomass separation'; this is one of the numerous added advantages of membrane technology. Additionally, membrane technology considerably improves biomass recovery, compared to traditional methods, because no cell damage occurs and less energy is used.

Polyether sulfone (PES), polyvinylidene fluoride (PVDF) and polysulfone (PSF) are often used in the manufacture of membranes because to their resilience towards chemical and physical disintegration. On the other hand, a key issue in the realm of membrane technology is membrane fouling. The hydrophobic barrier that exists between microbial cells and the membrane's constituent parts is the cause of this issue. To increase the membrane's hydrophilicity and lessen the fouling on its surface, three techniques have been considered: surface coating with nanomaterials, plasma treatment, and incorporation of nanoparticles. According to study investigations, NPs can greatly increase the hydrophilicity of membranes and significantly decrease membrane fouling. For instance, the surface modification (hydrophilicity) and antifouling capabilities of PSF hollow fiber membranes are enhanced by the addition of carbon nanotubes (CNT) and nano-titanium dioxide ($nTiO_2$). Madaeni and Ghaemi (2007) coated the polyvinyl layer of the RO membrane with $nTiO_2$, which, when exposed to UV light, reduced fouling through a self-cleaning mechanism. Due to their high surface area and hydrophilic traits, $nTiO_2$ photocatalysis was also used for the mitigation of air pollution. For this reason, the integration of these NPs with membranes reduces hydrophobicity and fouling. For the remediation of air pollution, photocatalytic $nTiO_2$ was used in addition to self-cleaning, manufactured "PVDF hollow fiber membranes with TiO_2 by phase inversion process on a custom designed single head spinning machine" in a different research study. Then, using an 'algal membrane bioreactor (A-MBR) for wastewater treatment', these membranes were put to the test. This was up to 75% efficient in the removal of phosphate and nitrogen from the water in this investigation. A decrease in membrane fouling and an improvement in PVDF's hydrophilic properties was also observed.

9.2 Aerobic Digestion with Nanoparticles

The method, generally used in traditional municipal wastewater treatment, is the activated sludge process. However, cleaning wastewater from industrial processes can be problematic due to the presence of harmful and hard-to-remove contaminants (Ali et al. 2021, Mehmood et al. 2022c). To ensure that the water is of good quality and that there are few chemical and biological impurities, many investigations on new technologies have been conducted. One method that can be employed is the use of zero valent iron (ZVI) (Chen et al. 2020), which can devalue organic contaminants and is consequently used in groundwater treatment to decompose chlorinated organic molecules. It has been suggested that the integration of biological water treatment techniques such as aerobic degradation with nano ZVI (nZVI) could realize the promise of large-scale bioremediation of organic chemicals prevalent in wastewater. According to the core idea behind such integration, half of the pollutants can be removed, with the remaining pollutants destroyed in the ensuing aerobic degradation phase. The underlying idea of such integration makes this possible. The micro- and nanoscale formation of bimetallic structures by cells plays a part in the first step's breakdown of contaminants. Following that, the products from nZVI take the charge of the biological breakdown of pollutants by aerobic bacteria. For instance, the granular Fe breaks down chlorine-containing solvents and petroleum hydrocarbons in the first stage, which is then followed by the dichlorination of chlorine-containing ethenes to reduce these organic pollutants. It must be noted that 'aerobic biodegradation' is not the same as 'biodegradation', even though the addition of DO in the second stage "facilitates the aerobic biodegradation of residual chlorinated compounds and any other hydrocarbons (Wang et al, 2017)." Similar to this, azo dyes and nitro-aromatic chemicals can be removed using nZVI in conjunction with biological procedures. Ma and Zhang (2008) carried out a smaller-scale investigation to look into the impact of nZVI on natural treatment. The results of this study showed that the effectiveness of reduction of BOD and COD was 96.5 percent and 86 percent respectively.

9.3 Enhancing the Effectiveness of Microbial Fuel Cells with Nanotechnology

Energy generation and water treatment from inorganic and organic chemicals, using, microorganisms as biocatalysts, are two benefits of adopting microbial fuel cell (MFC) technology for wastewater treatment, which has drawn attention of late. Because of these benefits, microbial fuel cell (MFC) technology for wastewater treatment has been developed. While microbes break down complex organic contaminants like acetate, butyrate and propionate into carbon dioxide (CO_2) and water (H_2O), the energy generated by MFCs has, in theory, the ability to lower the amount of electricity needed by any standard treatment system. However, the fact that this technique is less efficient than other fuel-cell-based technologies is impeding its commercialization. This method either needs to use a cheaper membrane and cathode catalyst, or its membrane maintenance costs need to be reduced; this technology will not be able to become commercially viable until then. For improving the effectiveness of MFCs, the choice of an appropriate material is also crucial. To modify the conductivity of MFCs, the electrode surface needs to improve since higher catalytic activity is needed for ORS. It is possible that the use of nanoparticles to improve certain properties is the best course of action.

Now, using inexpensive nanocomposite materials like nanostructured carbon is the main way to increase the efficiency of MFCs. These electrodes have a larger surface area, are highly conductive and are mechanically stable. Their electrochemical catalytic activity has also been noticeably improved. The application of carbon nanotubes (CNTs) in a membrane fuel cell (MFC) as electrodes at different chemical oxygen demand (COD) concentrations, including 100, 500, 1000 and 2000 mg/L showedthe CNT/Pt combination has the highest power generation across all COD media after CNT and Pt are applied independently. These results add to the body of evidence showing that the increased surface area of the electrode, due to CNT, boosts the catalytic activity of Pt. Carbon nanotube (CNT) electrodes have achieved noticeably improved outcomes due to their distinct structural characteristics and high conductivity. Consequently, the traditional cathode used in MFCs can be swapped with a CNT/Pt cathode to benefit from the latter's unique features. In order to improve microbial adhesion, carbon nanotubes have also been coated with a number of polymers, including polyaniline and polypyrene, to create a nanocomposite.

10. Current Perspectives and Future Prospects of NBT in the Treatment of Wastewater

Although NBT integration with other methods in water treatment studies have been conducted; yet, this area still lacks a generalized strategy, necessitating more research.

An additional requirement for research might be the creation of superior technological methods for enhancing nanomaterials. To bring down the overall costs, it is necessary to develop effective and trustworthy methods, with the aim of either anchoring nanoparticles on to the surfaces of the reactors or the different layers of filtration membranes, or separating and retaining suspended NPs. More specifically, strategies should be developed to reduce membrane fouling (using nanomaterial suspension), increase surface coating (possibly through nanoparticle surface functionalization), and impregnate nanoparticles (NPs) into filter packing materials like granular activated carbon or ion exchange resins. Any advancement in these fields would open the doors for the use of bionanotechnology in the current water treatment techniques. It might also be necessary to investigate cutting-edge separation methods or immobilization, like magnetic separation. For instance, extracting cadmium from water has been recently accomplished by mounting nanoparticles (NPs) on to magnetic supports such as nano magnetite

(Devi et al. 2017). The core of nano-magnetite can be combined with NPs of different characteristics to create a multifunctional nano-composite materials. Magnetic separation can be used to gather and recycle this material.

In order to effectively manage costs, economic scrutiny is required. One of the benefits of using conventional disinfectants is that they are readily available in nature and are relatively inexpensive. Bio-nanotechnology-based approaches towards water treatment will be able to constructively compete with more established technologies only if the systems that use NPs are cheaper, more efficient and viable. This calls for more research into the possible uses of low-cost nanomaterials for the generation of drinkable water and treatment of wastewater.

The validation of the methods used for wastewater treatment is another important aspect that must be considered to choose the appropriate treatment conditions and methodology. In particular, it is necessary to identify the fundamental experimental parameters that must be established in order to enable the types of bio-nanotechnology features among the various treatment approaches. For example, starting from a concentration of pollutants that is substantially lower than that of wastewater allows for the purification of drinkable water. Therefore, it is important to consider the concentration range of pollutants present in natural water resources, as well as the recommended upper and lower limits for drinkable water when assessing the performance of bionanotechnology. In order to assess the scalability of the treatment unit and the potential for further phases of post-treatment expansion, it is also necessary to conduct an appropriate assessment of the relevant kinetic rates.

It is necessary to optimize the treatment conditions to get consistent and desired results from laboratory investigations. Furthermore, it is recommended that these conditions be kept as close as possible to the characteristics of natural water resources. The pH of water is the most important factor that affects how effectively the procedure works when it is being carried out using NPs as the agents of treatment. Laboratory tests performed to remove as many contaminants as possible typically require a pH that is very close to acidic (below 5). However, in order for a method to be suitable for the treatment of drinking water, the pH of the experiment must be between 6 and 8. If the pH is lower than 6 or higher than 8, the method will not work.

It should be noted that considerable pH variations from the starting values may significantly impact the physicochemical properties of natural water, necessitating a further treatment step. This is something that should be kept in mind. Similarly, the simultaneous presence of the typical constituents found in natural water may compromise the effectiveness of NPs. Based on the nature of contaminants and the method by which NPs remove them, several cations (including Na^+, Mg^{2+} and Ca^{2+}) as well as some anions (including SiO_2^-, Cl^-, PO_4^{3-}, SO_4^{2-} and HCO_3^-) may serve as inhibitors. A few researchers have documented the effects of these variables, but there are very few open-access reports of their combined interactions.

Another apparent difficulty in developing these NPs technologies and the reported discrepancies among experimental research deploying NPs for water treatment is the apparent lack of common and meaningful indices for the judgment and comparison of treatment effectiveness. Examining the adsorption isotherms for experimental runs is the standard procedure. This is done in order to determine whether or not there was a decrease in the concentration of contaminants, which in turn validates the performance of the nanoparticles. In the case of removal of heavy metals, this is particularly important. A procedure like this exemplifies the elimination capacity (Q) of the nanomaterial, which can be quantified as the amount of pollutant or contaminant removed for each unit of NP used. However, most of the time, the maximum Q value, also known as Q max, is utilized. This value connects to the extremely high residual concentrations measured from where potable water demands are made.

11. Conclusion

Integrating biological processes into more sophisticated nanotechnology is a promising technique for treating and purifying wastewater. Numerous NPs, including nanofibers, CNTs, TiO_2 and ZVIs, have been carefully examined for use in wastewater treatment over the past 20 years.

Incorporating NPs into biological processes such as algal AMBR, MFC, and activated sludge has also resulted in increased outputs. The fact that activated sludge is used in these procedures proves this. One of the significant outcomes of adopting such integrated technologies in water treatment is increased efficiency concerning pollutant removal, dye decolorization, chemical oxygen demand (COD) and biological oxygen demand (BOD). NBT can improve wastewater treatment processes while posing fewer environmental risks. However, these techniques have some limitations, including the need for expensive equipment and specialized biological reagents for different pollutants (organic chemicals, dyes and nutrients), and require the ideal processing conditions for each microorganism.

In a not-too-distant future, applying nanotechnology-based water treatment techniques could be a viable way to address the associated difficulties, considering the effects on both human and environmental health. Because of this, the topic of distributed water treatment is witnessing an increase in attention, motivated by the worries concerning overworked water distribution systems.

The future study should emphasize the practicability, safety, scalability and operational cost factors of these systems. The treatment of drinking water dispensation networks and the reuse of wastewater could be revolutionized if inventions that address these issues overcome a significant fraction of the current limitations.

Acknowledgement

This work was supported by the Postdoctoral Research Project Fund of Hainan Province of China (2022-BH-03).

Abbreviations

ABM	Aquaporin base biomimetic membrane
A-MBR	Algal membrane bioreactor
CAP	Cellulose acetate phthalate
CNT	Carbon nanotube
COD	Chemical oxygen demand
DO	Dissolved oxygen
BOD	Biological oxygen demand
MFC	Microbial fuel cell
MMM	Mixed matrix membranes
MPNM	Marine polysaccharide-based nanomaterials
BMO	Biological manganese oxide
nMgO	Magnesium based nanoparticles
BFMO	Biogenic Fe-Mn Oxide
NP	Nanoparticle
nZVI	Zerovalent iron nanoparticles
PNP	Para nitrophenol
PSF	Polysulfone

References

Ali, I., Khan, D., Mehmood, T. and Khan, A. 2021. Removal of Uranium U (IV) from aqueous solution using acid treated spent tea leaves. Current Research in Green and Sustainable Chemistry 4: 100197.

Amin, M.T., Alazba, A.A. and Manzoor, U. 2014. A review of removal of pollutants from water/wastewater using different types of nanomaterials. Advances in Materials Science and Engineering 2014: 825910.

Ang, W.L., Mohammad, A.W., Hilal, N. and Leo, C.P. 2015. A review on the applicability of integrated/hybrid membrane processes in water treatment and desalination plants. Desalination 363: 2–18.

Aziz, N., Faraz, M., Pandey, R., Shakir, M., Fatma, T., Varma, A. et al. 2015. Facile algae-derived route to biogenic silver nanoparticles: Synthesis, antibacterial, and photocatalytic properties. Langmuir 31(42): 11605–11612.

Belessi, V., Lambropoulou, D., Konstantinou, I., Katsoulidis, A., Pomonis, P., Petridis, D. et al. 2007. Structure and photocatalytic performance of TiO2/clay nanocomposites for the degradation of dimethachlor. Applied Catalysis B: Environmental 73(3-4): 292–299.

Castro, L., Blázquez, M.L., González, F., Muñoz, J.A. and Ballester, A. 2018. Heavy metal adsorption using biogenic iron compounds. Hydrometallurgy 179: 44–51.

Celis, R., Adelino, M., Hermosín, M. and Cornejo, J. 2012. Montmorillonite–chitosan bionanocomposites as adsorbents of the herbicide clopyralid in aqueous solution and soil/water suspensions. Journal of Hazardous Materials 209: 67–76.

Chamjangali, M.A., Kouhestani, H., Masdarolomoor, F. and Daneshinejad, H. 2015. A voltammetric sensor based on the glassy carbon electrode modified with multi-walled carbon nanotube/poly (pyrocatechol violet)/bismuth film for determination of cadmium and lead as environmental pollutants. Sensors and Actuators B: Chemical 216: 384–393.

Chen, D.-w., Liu, C., Lu, J., Mehmood, T. and Ren, Y.-y. 2020. Enhanced phycocyanin and DON removal by the synergism of H2O2 and micro-sized ZVI: Optimization, performance, and mechanisms. Science of the Total Environment 738: 140134.

Choi, H., Zakersalehi, A., Al-Abed, S.R., Han, C. and Dionysiou, D.D. 2014. Nanostructured titanium oxide film-and membrane-based photocatalysis for water treatment. In: Nanotechnology Applications for Clean Water, Elsevier, pp. 123–132.

Das, S.K., Das, A.R. and Guha, A.K. 2009. Gold nanoparticles: Microbial synthesis and application in water hygiene management. Langmuir 25(14): 8192–8199.

Devi, V., Selvaraj, M., Selvam, P., Kumar, A.A., Sankar, S. and Dinakaran, K. 2017. Preparation and characterization of CNSR functionalized Fe3O4 magnetic nanoparticles: An efficient adsorbent for the removal of cadmium ion from water. Journal of Environmental Chemical Engineering 5(5): 4539–4546.

Dhillon, G.S., Kaur, S., Verma, M. and Brar, S.K. 2012. Biopolymer-based nanomaterials: Potential applications in bioremediation of contaminated wastewaters and soils. In: Comprehensive Analytical Chemistry, Vol. 59, Elsevier, pp. 91–129.

Dong, S., Feng, J., Fan, M., Pi, Y., Hu, L., Han, X. et al. 2015. Recent developments in heterogeneous photocatalytic water treatment using visible light-responsive photocatalysts: A review. Rsc Advances 5(19): 14610–14630.

Furgal, K.M., Meyer, R.L. and Bester, K. 2015. Removing selected steroid hormones, biocides and pharmaceuticals from water by means of biogenic manganese oxide nanoparticles in situ at ppb levels. Chemosphere 136: 321–326.

Gaurav, G.K., Mehmood, T., Cheng, L., Klemeš, J.J. and Shrivastava, D.K. 2020. Water hyacinth as a biomass: A review. Journal of Cleaner Production 277: 122214.

Gaurav, G.K., Mehmood, T., Kumar, M., Cheng, L., Sathishkumar, K., Kumar, A. et al. 2021. Review on polycyclic aromatic hydrocarbons (PAHs) migration from wastewater. Journal of Contaminant Hydrology 236: 103715.

Gupta, S.K., Behari, J. and Kesari, K.K. 2006. Low frequencies ultrasonic treatment of sludge. Asian Journal of Water, Environment and Pollution 3(2): 101–105.

Hassani, A., Khataee, A., Karaca, S., Karaca, M. and Kıranşan, M. 2015. Adsorption of two cationic textile dyes from water with modified nanoclay: A comparative study by using central composite design. Journal of Environmental Chemical Engineering 3(4): 2738–2749.

He, S., Guo, Z., Zhang, Y., Zhang, S., Wang, J. and Gu, N. 2007. Biosynthesis of gold nanoparticles using the bacteria Rhodopseudomonas capsulata. Materials Letters 61(18): 3984–3987.

Hennebel, T., Verhagen, P., Simoen, H., De Gusseme, B., Vlaeminck, S.E., Boon, N. et al. 2009. Remediation of trichloroethylene by bio-precipitated and encapsulated palladium nanoparticles in a fixed bed reactor. Chemosphere 76(9): 1221–1225.

Karri, R.R., Tanzifi, M., Yaraki, M.T. and Sahu, J. 2018. Optimization and modeling of methyl orange adsorption onto polyaniline nano-adsorbent through response surface methodology and differential evolution embedded neural network. Journal of Environmental Management 223: 517–529.

Khandel, P. and Shahi, S.K. 2016. Microbes mediated synthesis of metal nanoparticles: current status and future prospects. Int. J. Nanomater. Biostruct. 6(1): 1–24.

Klaus, T., Joerger, R., Olsson, E. and Granqvist, C.-G. 1999. Silver-based crystalline nanoparticles, microbially fabricated. Proceedings of the National Academy of Sciences 96(24): 13611–13614.

Kumari, S., Tyagi, M. and Jagadevan, S. 2019. Mechanistic removal of environmental contaminants using biogenic nano-materials. International Journal of Environmental Science and Technology 16: 7591–7606.

Kümmerer, K. 2009. Antibiotics in the aquatic environment — A review: Part I. Chemosphere 75(4): 417–434.

Leonard, P., Hearty, S., Brennan, J., Dunne, L., Quinn, J., Chakraborty, T. et al. 2003. Advances in biosensors for detection of pathogens in food and water. Enzyme and Microbial Technology 32(1): 3–13.

Li, J., Feng, H., Li, J., Jiang, J., Feng, Y., He, L. et al. 2015. Bimetallic Ag-Pd nanoparticles-decorated graphene oxide: a fascinating three-dimensional nanohybrid as an efficient electrochemical sensing platform for vanillin determination. Electrochimica Acta 176: 827–835.

Li, X., Xu, H., Chen, Z.-S. and Chen, G. 2011. Biosynthesis of nanoparticles by microorganisms and their applications. Journal of Nanomaterials 2011: 1–16.

Lingamdinne, L.P., Koduru, J.R., Chang, Y.-Y. and Karri, R.R. 2018. Process optimization and adsorption modeling of Pb (II) on nickel ferrite-reduced graphene oxide nano-composite. Journal of Molecular Liquids 250: 202–211.

Lirdprapamongkol, K., Warisnoicharoen, W., Soisuwan, S. and Svasti, J. 2010. Eco-friendly synthesis of fucoidan-stabilized gold nanoparticles. American Journal of Applied Sciences 7(8): 1038.

Ma, L. and Zhang, W.X. 2008. Enhanced biological treatment of industrial wastewater with bimetallic zero-valent iron.

Madaeni, S.S. and Ghaemi, N. 2007. Characterization of self-cleaning RO membranes coated with TiO2 particles under UV irradiation. Journal of Membrane Science 303: 221–233.

Mehmood, T., Ashraf, A., Peng, L., Shaz, M., Ahmad, S., Ahmad, S. et al. 2022a. Modern Aspects of phytoremediation of arsenic-contaminated soils. *In*: Global Arsenic Hazard: Ecotoxicology and Remediation, Springer, pp. 433–457.

Mehmood, T., Gaurav, G.K., Cheng, L., Klemeš, J.J., Usman, M., Bokhari, A. et al. 2021a. A review on plant-microbial interactions, functions, mechanisms and emerging trends in bioretention system to improve multi-contaminated stormwater treatment. Journal of Environmental Management 294: 113108.

Mehmood, T., Liu, C., Bibi, I., Ejaz, M., Ashraf, A., Haider, F.U. et al. 2022b. Recent developments in phosphate-assisted phytoremediation of potentially toxic metal (loid) s-contaminated soils. Assisted Phytoremediation, 345–370.

Mehmood, T., Liu, C., Niazi, N.K., Gaurav, G.K., Ashraf, A. and Bibi, I. 2021b. Compost-mediated arsenic phytoremediation, health risk assessment and economic feasibility using Zea mays L. in contrasting textured soils. International Journal of Phytoremediation 23(9): 899–910.

Mehmood, T., Mustafa, B., Mackenzie, K., Ali, W., Sabir, R.I., Anum, W. et al. 2022c. Recent developments in microplastic contaminated water treatment: Progress and prospects of carbon-based two-dimensional materials for membranes separation. Chemosphere 137704.

Mehmood, T., Zhu, T., Ahmad, I. and Li, X. 2020. Ambient PM2. 5 and PM10 bound PAHs in Islamabad, Pakistan: Concentration, source and health risk assessment. Chemosphere 257: 127187.

Menon, S., Rajeshkumar, S. and Kumar, V. 2017. A review on biogenic synthesis of gold nanoparticles, characterization, and its applications. Resource-Efficient Technologies 3(4): 516–527.

Mukherjee, P., Ahmad, A., Mandal, D., Senapati, S., Sainkar, S.R., Khan, M.I. et al. 2001. Fungus-mediated synthesis of silver nanoparticles and their immobilization in the mycelial matrix: A novel biological approach to nanoparticle synthesis. Nano Letters 1(10): 515–519.

Mustafa, B., Mehmood, T., Wang, Z., Chofreh, A.G., Shen, A., Yang, B. et al. 2022. Next-generation graphene oxide additives composite membranes for emerging organic micropollutants removal: Separation, adsorption and degradation. Chemosphere 136333.

Narayanan, K.B., Park, H.H. and Sakthivel, N. 2013. Extracellular synthesis of mycogenic silver nanoparticles by Cylindrocladium floridanum and its homogeneous catalytic degradation of 4-nitrophenol. Spectrochimica Acta Part A: Molecular and Biomolecular Spectroscopy 116: 485–490.

Narayanan, K.B. and Sakthivel, N. 2011. Synthesis and characterization of nano-gold composite using Cylindrocladium floridanum and its heterogeneous catalysis in the degradation of 4-nitrophenol. Journal of Hazardous Materials 189(1-2): 519–525.

Qu, X., Alvarez, P.J. and Li, Q. 2013a. Applications of nanotechnology in water and wastewater treatment. Water Research 47(12): 3931–3946.

Qu, X., Brame, J., Li, Q. and Alvarez, P.J. 2013b. Nanotechnology for a safe and sustainable water supply: Enabling integrated water treatment and reuse. Accounts of Chemical Research 46(3): 834–843.

Rai, M., Gade, A. and Yadav, A. 2011. Biogenic nanoparticles: an introduction to what they are, how they are synthesized and their applications. Metal Nanoparticles in Microbiology, 1–14.

Sabela, M.I., Mpanza, T., Kanchi, S., Sharma, D. and Bisetty, K. 2016. Electrochemical sensing platform amplified with a nanobiocomposite of L-phenylalanine ammonia-lyase enzyme for the detection of capsaicin. Biosensors and Bioelectronics 83: 45–53.

Saunders, J.A., Lee, M.-K., Dhakal, P., Ghandehari, S.S., Wilson, T., Billor, M.Z. et al. 2018. Bioremediation of arsenic-contaminated groundwater by sequestration of arsenic in biogenic pyrite. Applied Geochemistry 96: 233–243.

Sen, F., Boghossian, A.A., Sen, S., Ulissi, Z.W., Zhang, J. and Strano, M.S. 2012. Observation of oscillatory surface reactions of riboflavin, trolox, and singlet oxygen using single carbon nanotube fluorescence spectroscopy. ACS Nano 6(12): 10632–10645.

Sharma, D., Kanchi, S. and Bisetty, K. 2019. Biogenic synthesis of nanoparticles: A review. Arabian Journal of Chemistry 12(8): 3576–3600.

Sillanpää, M., Ncibi, M.C., Matilainen, A. and Vepsäläinen, M. 2018. Removal of natural organic matter in drinking water treatment by coagulation: A comprehensive review. Chemosphere 190: 54–71.

Tehrani, R.M., Ghadimi, H. and Ab Ghani, S. 2013. Electrochemical studies of two diphenols isomers at graphene nanosheet–poly (4-vinyl pyridine) composite modified electrode. Sensors and Actuators B: Chemical 177: 612–619.

Ternes, T.A., Joss, A. and Siegrist, H. 2004. Peer reviewed: scrutinizing pharmaceuticals and personal care products in wastewater treatment. Environmental Science and Technology 38(20): 392A–399A.

Thakkar, K.N., Mhatre, S.S. and Parikh, R.Y. 2010. Biological synthesis of metallic nanoparticles. Nanomedicine: Nanotechnology, Biology and Medicine 6(2): 257–262.

Wang, X., Zhang, D., Pan, X., Lee, D.-J., Al-Misned, F.A., Mortuza, M.G. et al. 2017. Aerobic and anaerobic biosynthesis of nano-selenium for remediation of mercury contaminated soil. Chemosphere 170: 266–273.

Xue, W., Hui, L., Min, W., Shu-Li, G., Yan, Z., Qing-Jiang, W. et al. 2013. Simultaneous electrochemical determination of sulphite and nitrite by a gold nanoparticle/graphene-chitosan modified electrode. Chinese Journal of Analytical Chemistry 41(8): 1232–1237.

Yadav, K., Singh, J., Gupta, N. and Kumar, V. 2017. A review of nanobioremediation technologies for environmental cleanup: A novel biological approach. J. Mater. Environ. Sci., 8(2): 740–757.

Yang, J., Dou, B., Yuan, R. and Xiang, Y. 2016. Proximity binding and metal ion-dependent DNAzyme cyclic amplification-integrated aptasensor for label-free and sensitive electrochemical detection of thrombin. Analytical Chemistry 88(16): 8218–8223.

Yuan, H., Sun, G., Peng, W., Ji, W., Chu, S., Liu, Q. et al. 2021. Thymine-functionalized gold nanoparticles (Au NPs) for a highly sensitive fiber-optic surface plasmon resonance mercury ion nanosensor. Nanomaterials 11(2): 397.

CHAPTER 6

Nanopharmaceuticals
A Step Towards Future Medicine
for Clinical Applications

Najm ur Rahman,[1,*] *Muhammad Riaz,*[1]
Abdul Sadiq[2] *and Rizwan Ahmad*[3]

1. Introduction

The concept of 'nanotechnology' was initially used by Norio Taniguchi in 1974 (Taniguchi 1974). It is now leading to step-by-step and attention-seeking innovations. This technology has marvelous prospects in pharmaceuticals for the improvement of therapeutic efficacy and reduction of toxicity due to the advanced system of (controlled, sustained, prolonged, enhanced, or actively or passively targeted or triggered) delivery of drugs. Based on many more major discoveries and developments in nanoscale materials science, the National Institute of Health (NIH) launched the National Nanotechnology Initiative in the year 2000 (Weissig et al. 2021). This program facilitated interdisciplinary research which led to the rise of nanomedicine as a new scientific discipline. According to NIH, nanotechnology is "the understanding and control of matter at dimensions between approximately 1 and 100 nanometers (nm), where unique phenomena enable novel applications." Drugs containing nanomaterials may be active ingredients or carriers. Emphasizing the

---
[1] Department of Pharmacy, Shaheed Benazir Bhutto University Sheringal Dir Upper Khyber Pakhtun Khwa, Pakistan.
[2] Department of Pharmacy, University of Malakand, Khyber Pakhtun Khwa, Pakistan.
[3] Natural Products and Alternative Medicines, College of Clinical Pharmacy, Imam Abdulrahman Bin Faisal University, Dammam, Saudi Arabia.
* Corresponding author: najm@sbbu.edu.pk

unique nanoscale properties and expanding the size scale beyond 100 nm, the FDA suggested that if a drug material or end product is engineered to show specific phenomena, these features are ascribable to its dimension(s), even if these are beyond the nanoscale range, i.e., up to one micrometer (1,000 nm), and can be considered as a nanomaterial (Weissig et al. 2021). In addition to nanomedicines used clinically, numerous nanomedicines are presently in different stages of pre-clinical and clinical development (Mitchell et al. 2021, Weissig and Guzman-Villanueva 2015). These contain products that exhibit potential therapeutic effects based on their unique nanoscale physicochemical properties such as superparamagnetism (exhibited by iron oxide nanoparticles) or improved bioavailability and disease specificity with lipid drug delivery carriers (such as those that have been used for FDA-approved COVID-19 vaccines) (Weissig et al. 2021).

2. Evolution of Nanomedicine

The evolution and development of nanopharmaceuticals attained significant consideration as a consequence of extensive restrictions and difficulties affecting conventional pharmaceuticals, classical formulations and conveyance systems (Farjadian et al. 2019). Reduced distribution to the target location is one of the inefficiencies of certain existing medications (Hare et al. 2017). That is why the distribution of medicaments in therapeutics is significantly important. In most conventional drug-delivery systems (CDDSs), the high release of drugs after administration and the increased frequency of administration lead to drug toxicity. The pharmaceutical industry also faces challenges in developing novel medications, due to the drug solubility issues with CDDSs, therefore affecting effectiveness (Farjadian et al. 2019, Hare et al. 2017). Furthermore, drug stability is also an issue of conventional pharmaceuticals as the CDDSs in some dosage forms are inappropriate to shield the active pharmaceutical ingredients (APIs) alongside the biological fluids in the body. Resolving all the preceding issues related to drug delivery will not only help in refining the value of life and the overall healthcare system, but also reduce healthcare costs. A decrease in healthcare expenses is one of the advantageous aspects of nanomedicines. Most countries are facing the pressure to lessen healthcare costs. Hence, extensive research is being conducted worldwide with the objective of enhancing nanomedicines, leading to the creation of products by domestic pharmaceutical industries. The aforementioned problems related to CDDSs can be effectively addressed by adopting nanotechnology in the development of nanopharmaceuticals.

Nanoparticles (NPs) in the manufacturing of medicines have significant benefits in comparison with CDDSs, like:

- Ease of therapeutic agent delivery to the targeted tissue, and minimizing the dose as well as the possible unwanted effects;
- Enhancing the stability and bioavailability of the APIs;
- Establishing better safety as well as efficacy;
- Control the release of drugs at a constant rate;
- Allowing passive targeting and accumulation of drugs through the EPR (enhanced permeability and retention) effect; and
- Cost-effectiveness, compared to conventional pharmaceuticals (Halwani 2022).

3. Characteristics of Nanoparticle Delivery Systems

Nanoparticles (1 to 1000 nm) are physically colloidal particles which are either entrapped, adsorbed or covalently attached to therapeutic agents. In drug development, the physicochemical properties of nanoparticles play a significant role to achieve controlled release features. Various factors that influence the efficacy of nanoparticles are stability, tolerability, simplicity in production and sterilization, etc. (Majuru and Oyewumi 2009). The important properties of nanoparticles considered in the development of nanomedicines are:

- High loading capacity and slow release ability of drugs.
- Potential to deliver a drug to the target site.
- Stability while passing through various barriers and reaching the target tissue/cell.
- Release of therapeutic agents from nanoparticles at an optimal rate according to the formulation design.
- Biocompatibility and biodegradable characteristics, as well as non-immunogenicity.
- Safe, inexpensive and commercially available components used in the formulation; avoiding toxic ingredients, as well as organic solvents.
- Simple, affordable and easy scale-up.
- Involved in various processes during the manufacturing of formulation, such as blending, drying, lyophilization, granulation, sterilization, compression, packaging, etc.
- Stale on storage.

The biocompatibility and biodegradability of nanocarriers should be good to ensure biosafety in clinical applications. Non-degradable nanocarriers such as metal oxides and carbon-based materials are toxic; this reduces their clinical applications. Lipid-based and polymer-based nanocarriers are useful for drug delivery due to their improved biocompatibility, lack of toxicity, and drug-loading capacity.

4. The Tendency of Nanomedicines Towards Clinical Applications

Nanomedicines (NMDs) are most often studied for the improvement of targeting specific sites of disease (i.e., site-specific drug delivery) and reducing the delivery of drugs to non-target tissues (Rizzo et al. 2013). Numerous nanomedicines in preclinical or clinical developmental stages, or in clinical use, are targeting various types of tumors and cancers (Hare et al. 2017). However, the use of nanomedicine therapy in non-cancerous circumstances has also increased recently, particularly for managing and exploring the biology of inflammatory diseases like asthma, inflammatory bowel disease, rheumatoid arthritis, diabetes, multiple sclerosis and neurodegenerative diseases (Hua et al. 2018).

4.1 Enhanced Permeability and Retention (EPR) Effect and Passive Accumulation of NMDs

The EPR of these nanomedicines will likely improve the favorable localization of NMDs in diseased tissues because of the intensified porosity of the vasculature that supplies such tissues. Besides the increased porosity of tumorous and inflamed blood vessels, the EPR effect also connects that solid tumors lack lymphatic drainage, which reduces the elimination of extravasated NMDs from the pathological site (Deamer 2010, Maeda et al. 2013). These properties accumulate NMDs at diseased sites; this is known as passive targeting. It is attained when nanoparticles with drug materials circulate for prolonged time spans in the bloodstream. Thus, the therapeutic efficacy of NMDs is expected to be higher than that of small molecules due to the EPR effect (Meel et al. 2016). The EPR effect was observed for the first time by Matsumura and Maeda in 1986 and has since been used predominantly in the evolution of NMs for the passive targeting of tumor, which leads the NMDs with suitable physicochemical properties and extended circulation half-life to accumulate in tumors over time (Danhier 2016). Thus, the EPR effect and the degree of passive targeting are highly dependent on the tumor's pathophysiology. It has been documented that EPR is a diverse phenomenon and can vary extremely within the same tumor type (Ojha et al. 2017). There is a positive correlation between tumor vascularization and passive targeting of NMs; for example, the first FDA-approved NMD, Doxil® (doxorubicin), proved superior effectiveness in ovarian cancer, compared to standard conventional therapies (Nichols and Bae 2014, Theek et al. 2014). In doxorubicin, encapsulated within

PEGylated liposomes, the circulation half-life is prolonged due to delays and reduction in the uptake and clearance by the reticuloendothelial system. Thus, NMDs accumulate in the tumor tissue due to increased leakiness of the tumor blood vessels, instead of healthy tissues which do not have such porous vessels. Moreover, the usage of pegylated liposomal doxorubicin reduces plasma peak levels of the free drug and significantly diminishes cardiotoxicity by avoiding the release of doxorubicin through the heart vasculature (Lyass et al. 2000).

4.2 Active Targeting of Nanomedicines

Active targeting involves ligands that are conjugated either physically or chemically on to the surface of NMDs in order to assist in the localization or targeting of diseased cells (Danhier 2016, Meel et al. 2016). Ligand-targeted NMDs can deliver therapeutic agents to specific sites, which selectively conveys specific receptors at the disease site (Hua 2013). For active targeting in cancer, the cellular targets are:

  i) Cancer cells which overexpress receptors for folate, transferrin, glycoproteins or epidermal growth factor;
 ii) Tumor endothelium which overexpresses endothelial growth factors, integrin receptors, or matrixes; and
iii) Stroma cells (e.g., fibroblasts, macrophages) (Coimbra et al. 2010, Danhier 2016, Kuijpers et al. 2010).

Ligand-targeted NMDs have enhanced therapeutic efficacy, despite no differences in the accumulation of NMDs between the targeted and the non-targeted tissues.

4.3 Triggered Release of Nanomedicines

This type of strategy is based on the stimulus-responsive release of drugs and is garnering much consideration from academia and the industry. NMDs in this class are engineered in such a way that the release of the drug enhances in response to exogenous or endogenous stimuli. NMDs responsive to endogenous stimuli exploit various factors linked with the microenvironment at the disease site, such as low pH, certain enzymes, or redox gradients in the environment of the tumor. While some exogenous stimuli such as light, temperature, magnetic field or ultrasound, etc. can induce the release of therapeutic compounds from exogenous-responsive NMDs. The use of an exogenous hyperthermic strategy or trigger to release therapeutic substances from NMDs (e.g., ThermoDox®) seems to be the most favorable method (Needham et al. 2000). The results of ThermoDox® usage in non-resectable hepatocellular carcinoma are superior to that of

Doxil® (Jang et al. 2016, Oude Blenke et al. 2013, Sawant and Torchilin 2012, Shi et al. 2017).

4.4 Nanomedicines Approved and in Clinical Trials

Numerous clinical-approved NMDs are available in the market, while a lot of them are in a developmental phase. NMDs in their clinical development already embrace the approved drugs on a variety of established drug delivery platforms like dendrimers, liposomes, polymeric micelles and inorganic nanoparticles (Sercombe et al. 2015, Torchilin 2006, Wagner et al. 2006). Despite the various targeted systems adopted in clinical trials and preclinical development, liposomes are the best and the first FDA-approved NMDs (Caster et al. 2017, Shi et al. 2017). Liposomes possess all the essential qualities to permit poorly soluble as well as highly toxic formulations/drugs, such as amphotericin B and paclitaxel (Caster et al. 2017, Min et al. 2015). Liposomes, discovered in 1965, were proposed as vehicles for the delivery of macromolecular as well as micromolecular drugs (Deamer 2010, Gregoriadis and Ryman 1971, Sessa and Weissmann 1968). After a long time of research, FDA approved the first liposomal nanomedicine Doxil® (doxorubicin) for the treatment of cancer (Allen and Cullis 2004). Liposomal formulations constitute the largest part of NMDs under clinical evaluation. Recently approved liposomal NMD, Vyxeos® (daunorubicin), exhibited better survival results, with manageable side effects, in phase III clinical trials on patients with therapy-related acute myeloid leukemia (Kim and Williams 2018).

5. Challenges in the Translation of Nanomedicines to Clinics

Translation of NMDs to clinics is an expensive as well as lengthy procedure. Conventional formulation technology is simpler than NMDs (Kumar Teli et al. 2010, Sainz et al. 2015, Tinkle et al. 2014). The main challenges in the clinical translation of NMDs include biological challenges, scale-up manufacturing, biocompatibility and safety, and government regulations (Narang et al. 2013, Sawant and Torchilin 2012, Zhang et al. 2008) (Figure 1). These factors limit the entry of NMs in the market, despite their therapeutic efficacy.

5.1 Biological Challenges

The association concerning biology and technology, including the effect of disease pathophysiology on the accumulation of nanomedicine, dissemination, retention and efficiency, along with the biopharmaceutical

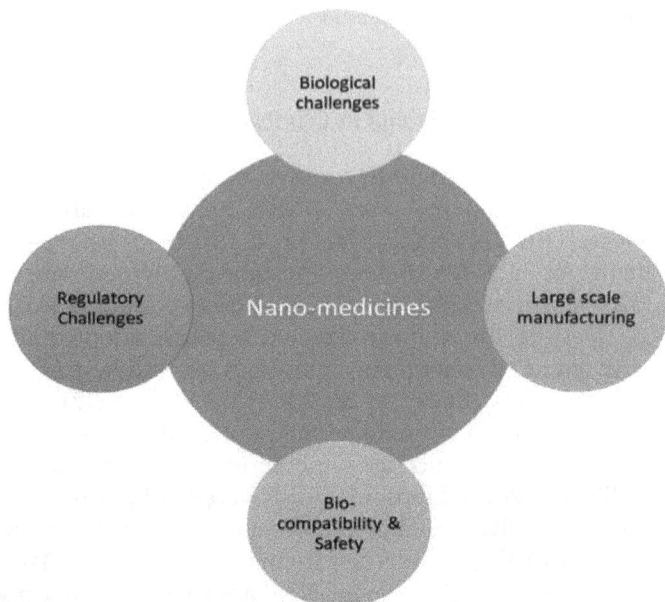

Figure 1: Challenges in the translation of nanomedicines to clinics.

relationship between delivery system features and *in vivo* response in animals versus humans, are vital factors for the translation of NMDs. Therefore, NMDs are designed and developed to exploit the pathophysiological changes in disease biology (Hare et al. 2017).

It is important to reflect on the correlation between pathophysiology and the diversity of diseases in humans, as well as the physicochemical characteristics of NMDs to improve targeting ill tissues and/or reducing amassing in healthy organs. A comprehensive and detailed evaluation of pharmacokinetics, biodistribution safety and therapeutic efficacy in appropriate experimental models relevant to human disease is essential in preclinical trials. The results of NMD use for a specific disease must be reproducible in different preclinical experimental models. Dissimilarities in the anatomy and physiology of experimental animals and humans should be considered while using different routes of administration. Preclinical trials of NMDs should be carried out under suitable randomization and the results should be evaluated against appropriate controls. The preclinical studies should be designed in such a way as to optimize the *in vivo* performance of NMDs, dosing plans and treatment combinations, as well as to understand the influence of disease severity and progression on the NMD's performance; this will affirm the degree of response of patients to NMD-based therapy (Hare et al. 2017, Hua et al. 2018).

More than 50% of the nanoformulations currently in the developmental stage or in clinical trials are targeting cancer therapy (Park 2017). Cancer

Therapy by NMDs is based on the EPR effect, although many reports have limited the EPR-mediated accumulation to certain tumor types (Maeda 2015). Tumors are also diverse like other diseases, and inter-patient as well as intra-patient inconsistencies are observed with the progression of the disease. The EPR effect in non-cancer cases has also been extensively studied for NMDs where leakiness occurs due to the porosity of inflamed blood vessels (e.g., atherosclerosis, rheumatoid arthritis, etc.) (Crielaard et al. 2012, Hua 2013, Maiseyeu et al. 2009, Metselaar et al. 2003). EPR is not the main contributing factor to a NMD's effectiveness as the latter cannot access all diseases, due to some biological barriers and the lack of the EPR effect in those diseases. The efficacy of NMDs is also subjected to the cellular uptake and release of drugs in the diseased tissues (Hare et al. 2017). Therefore, prediction of efficacy from preclinical data through a disease-driven approach will support the translation of NMDs to clinics.

5.2 Large-Scale Manufacturing

The manufacture of nanopharmaceuticals involves both quality and cost. Quality embraces the production procedure as well as the stability of the preparation. NMD development is currently facing potential challenges associated with: (i) lack of proper quality control, (ii) scalability difficulties, (iii) purity issue, (iv) elevated cost, (v) low yield, (vi) consistency, reproducibility, and storage stability, (vii) expertise and infrastructure deficiency, (viii) instability of compounds during the production procedure, and (ix) scarcity of investment (Hafner et al. 2014, Narang et al. 2013, Tinkle et al. 2014).

In order to clinically translate NMDs, it is essential to set specifications for the manufacturing of theses NMDs in large scalable quantities and high quality (Barz et al. 2015, Grainger 2013, Lammers 2013). Appropriate procedures for large-scale manufacturing of numerous basic NMD platforms have been effectively established (Jaafar-Maalej et al. 2012, Kraft et al. 2014). Issues arise when the NMD system becomes more complex (e.g., modifying the surface with ligands/coating, several targeting components and incorporating multiple therapeutic agents). Incorporation of several components into a single nanocarrier needs many steps, poses difficulties for scale-up manufacturing and raises the cost, making quality control and quality assurance more complex (Kumar Teli et al. 2010, Svenson 2012, Tinkle et al. 2014).

The validation and characterization of complex NMDs can be performed on a number of parameters (e.g., purity, size, morphology, distribution, charge, coating efficiency, encapsulation efficiency and density). Batch-to-batch dissimilarity of NMDs can alter the physicochemical features (e.g., size and polarity), pharmacokinetics (i.e., absorption, metabolism, distribution and excretion), and pharmacodynamics (e.g., activity and cellular interaction) (Barz et al. 2015, Kumar Teli et al. 2010, Tinkle et al. 2014).

5.3 Biocompatibility and Safety

A comprehensive toxicological study is also an essential factor for the clinical translation of NMDs to ascertain their safety for human use (Nyström and Fadeel 2012). Information on the effectiveness and toxicities of the free drug, the performance of diverse systems of NMDs, interaction with biological components, and the rate of drug release on-target and off-target predict the potential *in vivo* toxicities and side effects (Hare et al. 2017). The rational design of NMDs from the initial material assortment, manufacturing procedure optimization, and purification of the product is of fundamental significance to their clinical translation (Accomasso et al. 2018). Even though the safety of some common biodegradable polymers and phospholipids, increasing the complexity of NMDs due to the use of various synthetic compositions, ligands and coatings that significantly affect the biodistribution, biocompatibility and toxicological profile of nanomedicines (Allen and Cullis 2013, Narang et al. 2013, Sawant and Torchilin 2012, Tinkle et al. 2014). CARPA (complement activation-related pseudo-allergy) is an example of an immune adverse reaction caused by many NMDs (Jackman et al. 2016, Szebeni 2005, Szebeni and Storm 2015). The complement system (innate immune response) is involved in an array of immunological and inflammatory processes (Moghimi and Hunter 2001). CARPA is an immediate (non-IgE-mediated) hypersensitivity reaction that can cause facial swelling, anaphylaxis, facial flushing, headache, chills and cardiopulmonary distress (Szebeni 2005). This type of allergic reaction is usually managed either by reducing the infusion rate or stopping the therapy, as well as through the use of antiallergy medicines (e.g., corticosteroids, antihistamines and epinephrine) (Sercombe et al. 2015, Szebeni and Storm 2015). The unsusceptible response to NMD-based remedies can also alter the pharmacokinetics, mislay the efficiency and rise to serious life-threatening toxicities (Szebeni and Moghimi 2009, Szebeni and Storm 2015).

Detailed *in vitro* or *ex vivo* nanosafety assays are necessary before preclinical studies in animals. Standard protocols for *in vitro* should be followed using various cell culture models (i.e., liver, blood, lung, placenta, brain, gastrointestinal system) for the assessment of immunotoxicity, cytotoxicity and genotoxicity of NMDs (Accomasso et al. 2018). Although the existing testing methodologies are unsatisfactory for nanotoxicology assessments, numerous other approaches specific to NMDs are in the developmental stage, such as high-content screening, high-throughput screening techniques, and computational modeling (Accomasso et al. 2018, Dusinska et al. 2015, Nel et al. 2013, Oomen et al. 2014).

Short-term toxicity and long-term toxicity are also important to be studied in animal models because drug retention and circulation half-life increase with nanoencapsulation. Suitable analytical procedures are required for the monitoring of several features of the NMD drug delivery process, such as biodistribution, pharmacokinetics, accumulation at a target site and healing effectiveness (Kunjachan et al. 2015). Real-time imaging methods have assisted superior results in learning about the interactivity of NMDs with tissues and organs after *in-vivo* administration (Gaspar 2007, Kunjachan et al. 2015, Nyström and Fadeel 2012).

Furthermore, biocompatibility, inflammatory and immune-toxicological aspects should be evaluated, with functional results related to the mechanisms of tissue uptake as well as clearance. These variables are required to be exploited on the route of administration, dosage form, and dose of NMDs (that have never been used), for establishing harmless limits before clinical trials (Gaspar 2007, Nyström and Fadeel 2012).

5.4 *Regulatory Challenges*

Nanomedicines have momentous potential in the expansion of the pharmaceutical market and the improvement of health. However, numerous regulatory policies of the government in the manufacturing area, safety, quality control and patent protection pose as challenges for the commercialization of nanomedicines (Gaspar 2007). The unavailability of transparent safety and regulatory guidelines has badly affected the convenient and effectual clinical translation of NMDs. A suitable regulatory framework is required for helping in the evaluation of polymeric materials as well as polymer-based nanoformulations. Special regulatory protocols and standards for nanoparticles are required for bridging the medicine and medical device regulations. This should have the description of the complexity of NMDs, administration route, pharmacokinetics, safety profile, pharmacodynamics, clinical trial design, as well as the selection of patient (Gaspar and Duncan 2009, Sainz et al. 2015, Tinkle et al. 2014).

International regulatory standards and protocols for NMDs should be developed and established with key countries with invested interests. For this, a strong collaboration among the regulatory agencies, research, academia and industry is compulsory (Gaspar 2007, Hafner et al. 2014, Murday et al. 2009). This is of specific importance as very few organizations with specialty in manufacturing NMD formulations following the requirements for GMP are present in the world. These manufacturing organizations are required to be divided further based on their infrastructure and framework, and competencies of manufacturing definite nanomaterial platforms (e.g., dendrimers, polymeric nanoparticles, liposomes, etc.). Consequently, NMDs manufactured in these organizations will be marketed in many countries and uniform regulatory standards would be adopted (Hafner et al. 2014). The documentation and evaluation of manufacturing procedures for NMDs, incorporation of suitable industry standards for quality control, and preclusion of environmental issues should be completed in all aspects. Manufactured NMDs should also meet the standards generally adopted in the pharmaceutical industry, like manufacturing operations, purity, sterility, stability, etc. Advanced analytical methods and tools are required to be applied for the evaluation of physical parameters that affect the performance of NMDs, such as morphology, particle size, surface area, surface chemistry, surface coating, porosity, hydrophilicity, etc. (Gaspar 2007, Sainz et al. 2015, Tinkle et al. 2014). Regulatory authorities should work collectively to develop and establish suitable procedures for regulatory requirements and toxicity studies in order to ensure the efficiency and safety of the current as well as emerging NMDs.

6. Types of Nanocarrier Platforms

6.1 Carbon nanotubes: These nanotubes are made up of graphite sheets rolled into tubular or cylindrical structures, sealed either at one or both ends, ranging from 1–100 nm in length; they can be single-walled or multi-walled nanotubes (Iijima 1991). These tubes are used for drug encapsulation, given their size, surface features and vital physical parameters. DNA helix has a smaller diameter than single-walled nanotubes. Various methods like combustion procedures, electric arc discharge and chemical vapor deposition are used for the production of these nanotubes and fullerenes which target tissues and mitochondria. They also have antioxidant as well as antimicrobial activities (Saad et al. 2012).

6.2 Quantum dots: These nanocarriers (2–10 nm in size) are semi-conducting nanocrystals composed of an inorganic core and an organic shell, followed by a zinc sulfide coating which illuminates in light. The solubility of QDs in buffer solutions increases with the addition of cap (Iga et al. 2007). Several qualities like strong photo-stability, narrow emission, broad UV excitation and bright fluorescence have been ascribed to the extended tracking of intracellular processes, real-time monitoring and bio-imaging (Amiot et al. 2008). Biological performance and biomolecule detection, cell labeling, immunoassays, DNA hybridization, carriers for cancer therapy, creation of non-viral vectors for gene therapy, and transportation of biological and non-biological agents are some of the investigative and healing uses of QDs (Bailey et al. 2004).

6.3 Nanoshells: These are reformed prototypes composed of a silica core and coated with a metal, and have gained much attention for drug targeting (West and Halas 2000). These nanoshells have variable characteristics depending on the ratio between the shell and the core. The formulation of these nanostructures is easy and targeted based on their physical characteristics such as size and morphology. Due to the variation and differences in the morphology of metals, theses nanoshells are also used for the generation of a new system of different morphologies. To attain a suitable morphology, these shells are used for enclosing particles of specific shapes. These shells are inexpensive and smaller quantities of precious materials are required for the synthesis of these nanoshells (Kalele et al. 2006). Nanoshells perform different functions in different areas such as chemically stabilizing colloids, enhancing luminescence features, and drugs (Kherlopian et al. 2008).

6.4 Nanobubbles: These nanoparticles are formed at the confluence of lipophilic surfaces in liquids. They form micro bubbles when heated up to body temperature, and remain stable at room temperature. They emerge in supersaturated solutions as a consequence of gas nucleation at the hydrophobic surface, resulting in air gas trapping. These nanoparticles are of 4 types: bulk, plasmonic, oscillating and interfacial nanobubbles. These nanoparticles have been used for carrying anti-cancer drugs to cancerous tissues and the uptake by tumor cells has been reported to increase with the influence of ultrasound exposure (Gao et al. 2008, Klibanov 2006).

6.5 Paramagnetic nanoparticles: These are small particles (less than 100 nm in diameter) composed of magnetic elements; hence, they can be controlled under the influence of a magnetic field. These nanoparticles are classified according to their magnetic sensitivity. Paramagnetic nanoparticles have higher magnetic susceptibility than that of the typical contrast forms. These nanoparticles have a diagnostic potential and are used for treatment strategies as well. The targeting strategy of paramagnetic nanoparticles is effective for the identification of specific organs (Cuenca et al. 2006).

6.6 Liposomes: These are amphiphilic phospholipids, synthetic nanoparticles, made up of spherical double-layered vesicles, ranging in size from 50 nm to several micrometers. Liposomes have significant biocompatibility and biodegradability. In most clinical trials, liposomes are being used as drug vehicle nanosystems because of their characteristic properties to reduce the clearance of medication, along with decreasing toxicity (Torchilin 2005). Based on good pharmacokinetic characteristics, nanoscale modified liposomes are used for the transfer of proteins, siRNA, DNA, and for the treatment of cancer. However, some limitations of liposomes reduce their applications, like low loading capacity, quick release of a drug and the lack of adjustable drug release outlines (Nekkanti and Kalepu 2015). Due to their inability to penetrate cells, drugs are released in the extracellular fluid (Laverman et al. 2001). While using the oral route or the parenteral route, the structural integrity and stability in the environment can be achieved through surface modification (Lee et al. 2007). If the drug is coupled with the aqueous core of liposomes, using an ammonium sulfate gradient, the quick release of the drug can be restrained and the loss of the drug will thus be reduced (Gabizon et al. 2006).

6.7 Niosomes: These nanoparticles are molecular clusters of non-ionic surfactants formed by self-assembly in an aqueous layer. Owing to their unique architecture, niosomes can deliver lipophilic as well as lipophobic agents (Moghassemi and Hadjizadeh 2014). Niosomes are considered to be a replacement for liposomes because of their safety (non-toxic characteristics) and high stability. *In vivo*, niosomes have been observed to act like liposomes, for they extend the entrapped drug's circulation and improve distribution and metabolic stability. Besides the preparation technique, the characteristics of niosomes also rely upon the bilayer. Niosome nanosystems for drug delivery have a broad range in the entrapment of potent drugs (Malik et al. 2018), and anti-viral (Kazi et al. 2010) and anticancer medicines (Shah et al. 2019).

6.8 Dendrimers: These are multi-branched polymers. The size and shape of dendrimers depend upon the extent of branching. The cavities produced as a result of spherical branching inside dendrimers are responsible for the entrapment and delivery of a drug. Conjugation with other molecules takes place at the free ends of dendrimers (Moghimi et al. 2005). The three main regions of these nanocarriers are the core moiety, branching units, and the closely packed surface, which makes them advance in terms of drug delivery. Based on the structure, dendrimers have a wide range of delivery like vaccinations, genes, and medicines to particular tissues and cells. Solubilization, drug delivery, gene therapy, and immunoassay are some common uses of dendrimers (Huang and Wu 2018, Pedziwiatr-Werbicka et al. 2018).

6.9 Polymeric micelles: Polymeric micelles are self-assembled, lipophobic and lipophilic monomer units, i.e., amphiphilic copolymers. These are made up of a core of lipophilic blocks covered by a shell of lipophilic polymeric chains. Polymeric micelles have a delivery system superior to conventional systems. They improve the solubility of weakly water-soluble drugs, tissue permeation and accumulation at the target site, as well as stability, which results in enhancing the biodistribution bioavailability of the drug (Torchilin 2004).

6.10 Polymeric nanoparticles: These nanoparticles (1–1000 nm in size) are biocompatible and mostly biodegradable; therefore, they have gained the attention of researchers as a delivery system (Karlsson et al. 2018, Sahoo and Labhasetwar 2003). Based on their morphology, these nanocarriers are divided into two systems: matrix systems (nanospheres) and vesicular systems (nanocapsules). Recently, researchers have explored some synthetic polyesters for the adjustment of natural polymers. Chitosan is an example of a well-known natural polymer. Synthetic polymers have reduced the toxicity of many polymers (Bhatia 2016). Natural polymeric nanoparticles have greater efficiency and competence over traditional delivery systems. Yet, these nanoparticles also have some disadvantages such as degradation issues, poor reproducibility and potential antigenicity.

6.11 Nanocapsules: Nanocapsules and nanospheres are two types of polymeric nanoparticles. Nanocapsules are carriers containing drugs in the core encapsulated by a polymeric membrane, while nanospheres disseminate drugs through the polymeric matrix (Mora-Huertas et al. 2010). The drug may be dissolved, encapsulated or entrapped inside or throughout the matrix. Through this delivery system, drugs are used for the treatment of cancer and other diseases (Kayser et al. 2005).

6.12 Solid lipid nanoparticles: These nanoparticles are produced from solid lipids matrices, with a mean size of 40–1000 nm. They are stabilized by the use of a surfactant layer. These nanoparticles are used as an alternative to other systems by overcoming their limitations (Kayser et al. 2005). These nanoparticles have characteristics superior to other drug delivery systems, like tolerability, bioavailability, biodegradability, and target specificity (Cavalli et al. 2002, Yang et al. 1999). Extensive research has been conducted on these nanoparticles for a variety of lipophilic and hydrophilic drugs.

6.13 Nanoemulsions: Emulsion is a dispersed system of immiscible liquids where one liquid is disseminated in the form of small droplets inside the other. Recently, attention towards nanoemulsions and self-emulsified drug delivery systems (SEDDS) has been enhanced for the bioavailability of drugs with low solubility in water (Singh and Vingkar 2008). The oral bioavailability of weakly aqueous soluble drugs increases through this system. Moreover, the surface tension between the small oil droplets and the aqueous medium of the GIT reduces, resulting in a more uniform and extensive distribution of the drug in the gut (Cai et al. 2010, Mazayen et al. 2022).

7. Significant Role in the Healthcare System

A. Nanoparticles in the Treatment of Infectious Diseases

The emerging and re-emerging diseases caused by various infectious agents such as bacteria, viruses, fungi and parasites, have severely affected the public health globally and are considered to claim the highest number of casualties across the world. The transmission of these infections varies, depending on microorganisms, as some infections are highly transmissible and of virulent nature. Drug resistance is a major disadvantage and a public health threat associated with conventional systems in the treatment of these infections. The development of antibody-based therapeutics and metal-based nanoparticles has changed the era of these diseases (Casadevall 1996, Saylor et al. 2009) (Figure 2).

Metal-Based Nanoparticles with Antibacterial Activity

The small size and high surface area of these nanoparticles influence their cellular uptake and enhance their cellular interactions (Arvizo et al. 2012). Nanoparticles, by producing reactive oxygen species (ROS), destroy the bacteria and, by binding to RNA or DNA, hamper the microbial replication processes (Aderibigbe 2017, Zain et al. 2014). Researchers have reported different types of metal-based nanoparticles, but the mechanism of some is still unknown. Some well-known metal-based nanoparticles with bactericidal activity are described in the succeeding sections.

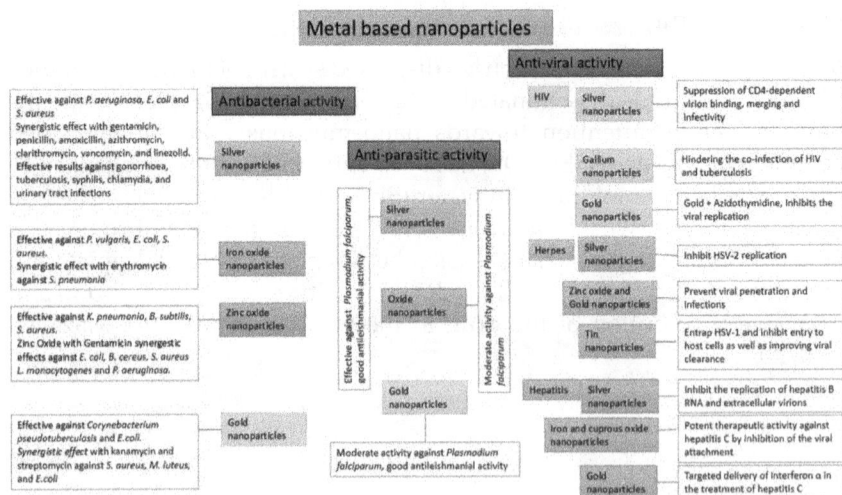

Figure 2: The role of metal-based nanoparticles against infectious diseases.

1. Silver nanoparticles

These nanoparticles penetrate the bacterial cell wall, interact with vital enzymes, and destroy membrane permeability, thus resulting in bacterial death. Researchers have proposed that silver nanoparticles interact with the phosphorus and sulfur atoms of bacterial DNA, thus preventing its replication, thereby causing cell death. Silver nanoparticles also inhibit signal transduction by altering the phosphotyrosine profile, resulting in cell growth inhibition (Morones et al. 2005, Prabhu and Poulose 2012). Silver nanoparticles are reported in various sizes and shapes (like nanoplates, nanospheres, nanocubes, triangular nanoparticles, nanorods, hexagonal nanoparticles, nanoprisms and nanowires). The bactericidal action of these nanoparticles is ascribed to their size, shape and concentration; for instance, smaller nanoparticles have a large surface area and are more effective against *P. aeruginosa*, compared to larger silver nanoparticles; similarly, spherical nanoparticles have superior results than triangular nanoparticles (Raza et al. 2016). Hexagonal silver nanoparticles are more efficacious against *E. coli* and *S. aureus*, compared to nanoplates and nanorods (Romaniuk and Cegelski 2015). Silver nanoparticles, in combination with various antibiotics like gentamicin, penicillin, amoxicillin, azithromycin, clarithromycin, vancomycin and linezolid, also show synergistic results against various bacterial infections (Smekalova et al. 2016). *In vitro* and *in vivo* findings reveal that silver nanoparticles are effective for the therapy of various bacterial infections such as gonorrhea, tuberculosis, syphilis, chlamydia and urinary tract infections.

2. Iron Oxide Nanoparticles

These nanoparticles are more effective against Gram-positive bacteria rather than Gram-negative bacteria. The bactericidal activity of these nanoparticles is linked to the generation of oxidative stress by reactive oxygen species. This oxidative stress leads to DNA damage, protein degradation through oxidation, and disruption of cell membranes through electrostatic interaction of the bacteria (Rafi et al. 2015). The antibacterial activity of these nanoparticles depends on their preparation procedures, concentration, charge (positive/negative) and shape. Iron oxide nanoparticles have good antibacterial action against *P. vulgaris, E. coli* and *S. aureus*, while synergistic effects in combination with erythromycin have been observed against *S. pneumonia* (Behera et al. 2012, Gupta and Gupta 2005, Ismail et al. 2015).

Other metal-based nanoparticles have also been reported for antibacterial activity. Zinc oxide nanoparticles with good antibacterial activity against *Kl. pneumonia, B. subtilis* and *Staphylococcus aureus* have been reported in literature (Hsueh et al. 2015, Narasimha et al. 2014, Reddy et al. 2014). Some researchers have reported that zinc oxide nanoparticles show dose-dependent antibacterial activity and are more effective against *S. aureus* (Gram-positive) than *E. coli* (Gram-negative). Zinc oxide nanoparticles also synergize with the antibacterial activity of gentamicin against various strains like *Escherichia coli, B. cereus, L. monocytogenes, P. aeruginosa* and *S. aureus* (Voicu et al. 2013). Similarly, aluminum oxide-based nanoparticles also have a potential against both Gram-positive and Gram-negative bacteria, but the inhibition is not of significant level (Brintha and Ajitha 2016, Prashanth et al. 2015, Sadiq et al. 2009).

Gold-based nanoparticles have also been reported to have antibacterial activity against *Corynebacterium pseudotuberculosis* and *E. coli* (Mohamed et al. 2017, Zhou et al. 2012). Gold nanoparticles in combination with kanamycin and streptomycin have shown synergistic effects against *S. aureus, M. luteus* and *E. coli* (Saha et al. 2007).

Metal-based Nanoparticles for the Treatment of Viral Infections

The management of viral infections via the conventional system faces major issues because of the drug resistance developed due to viral replication and prolonged contact with the drug, which causes drug toxicities and even more complications (Strasfeld and Chou 2010). Therefore, an effective drug system is required to treat these infections. Various effective metal-based nanoparticles are reported for the management of viral infections.

HIV

HIV/AIDS is a severe viral infection that has affected more than 38 million people globally. Silver, gold and gallium nanoparticles have been found to diminish the growth as well as the replication of HIV. Silver nanoparticles disrupt the replication of viruses and the cycle of HIV-1 by binding to the gp 120 glycoprotein. This binding leads to the suppression CD4-dependent fusion and infectivity of the virions (Strasfeld and Chou 2010). Gallium nanoparticles have also been reported for hindering the co-infection of HIV and tuberculosis (Choi et al. 2017). Gold nanoparticles enhance the effectiveness of anti-HIV drugs. Gold nanoparticles in combination with azidothymidine inhibit viral replication, and an effective antiviral has been reported in conjugation with peptide triazole as well (Bastian et al. 2015, Kesarkar et al. 2015).

Herpes

Herpes is another viral disease attributed to the herpes simplex virus. This virus causes invasive cervical carcinoma (Vonka et al. 1984). The conventional system is unable to eradicate the virus which can produce complications like encephalitis. Reports reveal that tin nanoparticles can entrap HSV-1 and inhibit their entry to host cells, thus reducing replication and improving viral clearance (Trigilio et al. 2012). Silver nanoparticles inhibit HSV-2 replication and infection when modified with tannic acid. Also, zinc oxide and gold nanoparticles prevent viral penetration, resulting in the prevention of infections (Mishra et al. 2011, Sarid et al. 2014).

These metal nanoparticles have also been found to be potential therapeutics in the management of other viral infections; for example, silver nanoparticles inhibit the replication of hepatitis B RNA and extracellular virions (Lu et al. 2008). Iron oxide and cuprous oxide nanoparticles have been reported to have potent therapeutic activity against hepatitis C, through the inhibition of viral attachment (Hang et al. 2015, Ryoo et al. 2012), while gold nanoparticles have been reported for the targeted delivery of interferon α in the treatment of hepatitis C (Lee et al. 2012). The effectiveness of silver and gold nanoparticles has also been reported against the influenza virus (Feng et al. 2013).

Metal-based Nanoparticles for the Treatment of Parasitic Infections

Very few reports are available on metal-based nanoparticles for the management of parasitic infections. In the evaluation of antimalarial activity, silver nanoparticles suppressed the growth of *Plasmodium falciparum*, while oxide nanoparticles and gold nanoparticles were reported to have moderate activity (Jacob Inbaneson and Ravikumar 2013, Karthik et al. 2013). Silver nanoparticles have been reported to have good antileishmanial activity, while gold nanoparticles show activity against both wild as well as drug-resistant stains (Halder et al. 2017, Zahir et al. 2015).

B. Nanoparticles/Nanomedicines for the Treatment of Cancer

Treatment of cancer using the conventional system is a big challenge due to the severe side effects, multi-drug resistance, lack of targeting ability and insufficient delivery of drug to the target site. The small size of nanoparticles not only increases the concentration at the target site, but also enhances the targeting potential of multiple therapeutic agents through conjugation on their surface, and reduces the adverse effects. Numerous FDA approved nanomedicines for the treatment of cancer exist since 1995 (Dawidczyk et al. 2014). Some commercially available nanomedicines have been discussed in the following paragraphs.

Doxil® (Unilamellar PEGylated-Liposomal Doxorubicin), also marketed as Caelyx®, Evacet® and Lipo-Dox®, was the first liposomal formulation approved by the FDA, in 1995, for the therapy of Kaposi's sarcoma linked with AIDS (acquired immune deficiency syndrome), multiple myeloma and ovarian cancer (Duggan and Keating 2011). The nano size (80–90 nm) of liposome and PEGylation increases the circulation half-life, bioavailability and drug amassing at the tumor site (Zhao et al. 2018). Doxil passively targets the cancer cells by intercalating with DNA molecules, which results in the destruction of the topoisomerase enzyme, repairing of DNA, and production of free radicals and reactive oxygen species which destroy cellular membranes (López 2000).

Myocet,® a non-PEGylated doxorubicin liposome (150 to 250 nm), was approved in 2000 for the management of metastatic breast cancer. The size of this nanomedicine enhances its retention in the tumor cells and minimizes the adverse effects associated with doxorubicin (Leonard et al. 2009).

DaunoXome® (daunorubicin citrate), a liposomal formulation, was approved commercially in 1996 for the chemotherapy of HIV-linked Kaposi's sarcoma and acute myeloid leukemia. This formulation, encapsulating daunorubicin in the aqueous core, improves the stability and protection metabolism of the liposomes, converting active daunorubicin to an inactive moiety (daunorubicinol), thus reducing cardiotoxicity as well (Fassas and Anagnostopoulos 2005).

Abraxane® is an albumin-bound/protein-nanotechnology-based formulation (paclitaxel NPs, size 130 nm) approved in 2005 for the chemotherapy of breast cancer, in 2012 for lung cancer, and in 2013 for pancreatic adenocarcinoma (in combination with carboplatin) (Farjadian et al. 2019). Paclitaxel inhibits cellular mitosis by averting microtubule depolymerization, hence reducing the agglomeration of tubulin. Comparative studies have shown abraxane with better results than conventional Taxol (Gradishar et al. 2005).

Onivyde,® known as MM-398 or PEP02, is a liposomal irinotecan nanoformulation (particle size of 110 nm) approved by the FDA, in 2015, for the chemotherapy of metastatic adenocarcinoma of the pancreas. It also has synergistic effects with other anticancer drugs. The advantages of this nanoliposomal formulation of irinotecan are improved efficacy, biodistribution, enhanced circulation time, passive targeting of tumors, accretion at tumor sites and reduced unwanted effects. However, abdominal pain, diarrhea, vomiting and alopecia are the reported negative effects (Zhang 2016).

DepoCyt® is a controlled-release, liposomal (multivesicular) formulation of cytarabine, produced by using DepofoamTM technology in which the therapeutic agent cytarabine is encapsulated into multivesicular particles (Gadekar et al. 2021). It was introduced under the accelerated approval regulations in 1999 and was fully approved in 2007 for the therapy of lymphomatous meningitis. The particle size herein ranges from 10–100 μm, and these particles are capable of entrapping large quantities of drugs. Liposomal cytarabine is administered intrathecally into the cerebrospinal fluid. The half-life of DepoCyt® is 40 times longer than that of conventional cytarabine (Kim et al. 1993).

Marqibo® is a liposome-based vincristine sulfate formulation, approved in 2012 for the treatment of Philadelphia chromosome-negative lymphoblastic leukemia (Silverman and Deitcher 2013). Vincristine binds to tubulin and inhibits cell division. Marqibo (vincristine encapsulated in sphingomyelin) shows increased circulation time in the blood (higher AUC and delayed clearance), superior release profile and better accretion in tumors, compared to conventional systems (Boman et al. 1994). The significant side effects here are constipation, fatigue, nausea, diarrhea and insomnia (Thomas et al. 2006).

Genexol-PM® is a paclitaxel containing an IV injection of lyophilized polymeric micelle, launched in Korea, in 2007, for the chemotherapy of breast cancer and lung cancer. The micelle (size 20–50 nm) shows increased circulation time (due to a hydrophilic shell), enhanced solubility and efficacy, higher accumulation in tumor tissue, and reduced toxicity than the conventional Taxol (Werner et al. 2013).

Vyxeos® is a combination of cytarabine and daunorubicin encapsulated in liposomes, approved in 2017, for the chemotherapy of adults affected by acute myeloid leukemia. Both the drugs affect the synthesis of DNA and generate free radicals which destroy DNA (Crain 2018). Vyxeos prolongs the bioavailability while decreasing dissemination to ordinary tissues (Farjadian et al. 2019). Its side effects include hypersensitivity reactions, cardiotoxicity and tissue necrosis (Crain 2018).

C. Nanoparticles/Nanomedicines for the Treatment of Fungal Infections

Dermal diseases are mainly caused by fungal infections. Resistance to fungal infections is a major drawback of conventional treatments. Nanomedicines play a vital role in this case; for instance, it increases absorption, reduces the formation of biofilms, decreases the outflow of drugs from the cell and enhances the accretion of drugs at the infection site in order to overcome the drawback of conventional systems (Voltan et al. 2016). Some commercially available nanomedicines have been described hereunder.

AmBisome® is the liposomal formulation of Amphotericin B (AmB) or L-AmB (size 60–70 nm), approved in 1997 for the treatment of systemic fungal infections. It does not produce resistance, compared to other antifungal drugs. Liposomal AmBisome enhances the pharmacokinetics and the stability of the drug during circulation, thus reducing accumulation in uninfected tissues and decreasing toxicity; hence, it is safer than Fungizone.

Amphotec® is a lyophilized parenteral AmB cholesteryl sulfate lipid complex approved, in 1996, for the treatment of fungal infections and leishmaniasis (Clemons and Stevens 1998). AmB and Cholesteryl sulfate form a stable complex. This results in a microscopic colloidal system with higher stability, reduced hemolysis and acute toxicity in an aqueous solution.

D. Nanoparticles/Nanomedicines for the Treatment of Neurological Disorders

The inability of conventional systems to pass through the blood-brain barrier is the main constraining factor for the therapy of neurological diseases. NPs can pass this barrier and deliver the therapeutic agent to a specific target. Several commercially available nanomedicines for the management of neurological disorders have been described in the succeeding paragraphs.

Avinza® is an oral (hard gelatin capsule) formulation comprised of instant release (10%) and controlled release (90%) morphine sulfate (1–2 mm beads), approved by the FDA in 2002 (Bobo et al. 2016). Avinza® releases 10% of morphine in the gastrointestinal fluid, which is immediately absorbed; this triggers its activity. The swellable polymer coating is used for the formation of sustained-release beads. Fumaric acid facilitates the swelling of the polymer, resulting in the formation of pores on the polymeric; the drug is released through these pores in a controlled pattern. Avinza® produces a steady and predictable pain relief level (Caldwell 2004).

Ritalin LA,® a psychostimulant, nanocrystalline, sustained-release oral capsule, was approved in 2002 for the treatment of ADHD (attention-deficit/hyperactivity disorder) in children below six years of age. Ritalin LA® contains a therapeutic drug, methylphenidate hydrochloride; 50% of the dose is immediately released, while the remaining (50%) exhibits extended action (Biederman et al. 2003, Lyseng-Williamson and Keating 2002). Focalin XR® (d-threo-enantiomer of methylphenidate) also has the same efficacy in the management of ADHD (Gao et al. 2013).

Invega Sustenna® is a long-acting, extended-release, injectable nanocrystalline formulation of paliperidone palmitate, approved in 2009 for the treatment of craziness. It has the advantages of 100% drug loading and improved pharmacokinetics and biodistribution (Lu et al. 2017). It causes depot formation after intramuscular injection and the drug is released slowly, resulting an extended half-life of the formulation (Abruzzo et al. 2019, Park et al. 2013).

E. Nanoparticles/Nanomedicines for the Treatment of Ocular Disorders

Nanoparticles have broad applications in the delivery of lipophilic as well as hydrophilic drugs. Their penetration through the corneal barrier not only increases the efficacy of the drug but also improves the management of ocular diseases (Weng et al. 2017). Some commercially available nanomedicines used in the management of ocular diseases have been discussed below.

Visudyne® is a liposomal nanoformulation of a photosensitizer (verteporfin), approved in 2001. Verteporfin is a benzoporphyrin derivative used as a photosensitizer in Photodynamic Therapy PDT. The absorption peak of verteporfin is at 692 nm, which, on irradiation, causes cytotoxicity as a consequence of photochemical reactions. Their particle size ranges from 150–300 nm. The 100% encapsulation of the photosensitizer due to its lipophilicity results in better accretion of the photosensitizer in the tumor, compared to hydrophilic preparations. Visudyne® shows excellent results in optical therapy, albeit with several undesirable effects such as dryness, redness, slight vision changes and swelling of eyes (Nayak and Misra 2018).

Restasis® is a cyclosporin-containing ophthalmic anionic nanoemulsion approved in 2002. In 2003, it was launched by Allergan in the US. Restasis® is a nanoemulsion (oil in water) in which cyclosporine is dissolved in castor oil with an emulsifying agent polysorbate 80. EMA, in 2015, allowed Novasorb (cationic nanoemulsion) for the therapy of severe keratitis. The cationic nanoemulsion exhibited high bioavailability and half-life of cyclosporine. Also, there was better ocular residence time with increased ocular cyclosporine. The clearance rate for Novasorb is 57-times lesser than that of Restasis (Lallemand et al. 2017).

F. Nanoparticles/Nanomedicines for the Treatment of Cardiovascular Diseases

Nanotechnology-based medicines play a vital role in the management of various cardiac disorders such as cardiac arrhythmia, infarction and atherosclerosis. A well-known nanomedicine for such cases is Tricor,® a nanocrystalline formulation (tablet) of fenofibrate, which got approval in 2004 for the management of hypercholesterolemia or mixed dyslipidemia in adult patients. This drug is used in adjunctive therapy with another lipid-lowering diet or another drug. The bioavailability of the nanocrystal formulation is independent of the meal, unlike conventional fenofibrate (Joseph and Singhvi 2019). The nano-ionization of the formulation results in better solubility, high bioavailability, and increased muco-adhesiveness to the intestine (Jarvis et al. 2019). Similarly, Triglide® is another fenofibrate formulation that was developed in 2005.

G. Nanoparticles/Nanomedicines for the Treatment of Blood Disorders

Several nanoparticles/nanomedicines used in the management of blood disorders are Feraheme,® also known as Ferumoxytol, a semi-synthetic iron oxide nanoformulation (30 nm in size), approved in 2009 for the treatment of anemia associated with chronic kidney disease (Bullivant et al. 2013). It is composed of a super-paramagnetic iron oxide core coated with poly glucose sorbitol carboxymethyl ether. The nanoparticles coated with carbohydrates facilitate the release of drug, especially when it comes in contact with the reticuloendothelial system of the liver. Feraheme® has a superior safety profile than other similar iron formulations. After administration, Feraheme® exhibits an immediate and effective response for the management of anemia (Hassan et al. 2017).

Ferinject,® or Injectafer,® is the non-dextran, intravenous iron nanoformulation, approved by FDA in 2013 for the management of iron deficiency anemia (Kulnigg et al. 2008). Like Feraheme,® it is composed of a polynuclear iron (III)-oxyhydroxide core covered with carboxymaltose, with an average particle size of 23 nm. This novel formulation permits administration up to 1000 mg of iron replenishment dose in a short period of time (15 minutes), with reduced risk of undesirable effects. The half-life (7–12 hours) of Injectafer® reduces the frequency of doses (Moore et al. 2011, Pai 2017).

Monofer® is an iron isomaltoside 1000 injectable nanoparticle formulation, introduced in 2010 for the treatment of iron deficiency anemia (Reinisch et al. 2013). Monofer® is an extended slow-release formulation composed of carbohydrate and iron molecules embedded in a matrix (Jahn et al. 2011). This formulation permits flexible dosing and reduces toxicities as a result of free iron and immunological reactions (Kalra 2011, Kalra and Bhandari 2016). Besides Monofer,® Injectafer® and Feraheme,® several similar nanoformulations with different compositions of carbohydrate shell, are available for the treatment of iron deficit anemia, such as Ferrlecit® (loosely associated with gluconate), Venofer® (sucrose), InFed® (dextran polysaccharide shell), etc. (Jahn et al. 2011). However, some toxicities have been reported with dextran-containing nanoformulations such as DexFerrum® and CosmoFer® (Fütterer et al. 2013).

Mircera® is the polymer-protein conjugate approved in 2007 for the treatment of anemia associated with kidney diseases (Curran and

McCormack 2008). Mircera® is an erythropoietin receptor activating agent produced by the PEGylation of erythropoietin beta (McGahan 2008). The polymer protein is made by the conjugation of methoxyPEG (mPEG) with protein. The conjugated product has an extended half-life (130 hours), compared to the conventional erythropoietin half-life of 4–10 hours (Reynaldo et al. 2018).

Neulasta® (pegfilgrastim) is another polymer protein conjugate synthesized by the PEGylation of filgrastim, approved in 2002 for the treatment of febrile neutropenic patients. Neulasta® exhibits an increased hydrodynamic volume, along with a reduced rate of clearance (Piedmonte and Treuheit 2008). Filgrastim has a shorter half-life (3.5 hours) and requires daily administration till the retrieval of neutrophils, whereas pegfilgrastin has a prolonged half-life and needs one dose per treatment cycle (Molineux 2004).

Similarly, Pegasys® (PEGinterferon alfa-2a) and PEGintron® (PEGinterferon alfa-2b) got FDA approval in 2002 and 2001 respectively, for the treatment of Hepatitis B and C (Weissig et al. 2014). PEGylation increases the solubility and half-life of the protein, while decreasing the renal excretion of interferons. PEGylation also reduces the dosing frequency to once in 12 weeks, compared to 3 times a week for free interferons. Pegasys® in combination with ribavirin and lamivudine yields better results for Hepatitis C and chronic Hepatitis B respectively (Reddy et al. 2002).

H. Nanoparticles/Nanomedicines for the Treatment of Immune and Hormonal Disorders

The human body has an innate defence system called the immune system. When an abnormality occurs in the immune system, it leads to immunological disorders like allergy, psoriasis, asthma, rheumatoid arthritis, inflammatory bowel disease, multiple sclerosis (MS), type-1 diabetes mellitus, systemic lupus erythematosus, etc. On the other hand, hormones are chemical messengers produced in the body for controlling various vital processes such as metabolism and reproduction, etc.; any imbalance with hormones causes various abnormalities. Some examples of approved nanomedicines available in the market for the treatment of immune and hormonal disorders have been discussed below.

Copaxone® is a synthetic polypeptide colloidal preparation composed of glatiramer acetate (active) and a copolymer of amino acids (L-lysine, L-alanine, L-glutamic acid and L-tyrosine), approved in 1996 as a subcutaneous injection to treat multiple sclerosis (Melamed-Gal et al. 2018). Copaxone® exerts immunomodulatory action on innate as well as acquired immunity. It is effective and safe for the treatment of multiple sclerosis.

Estrasorb® is a micellar topical lotion (mixture of water, oil, surfactant, alcohol and 17β-estradiol), approved in 2003 for the management of menopause-associated vasomotor symptoms such as night sweats, disturbed sleep patterns, hot flushes, lethargy, fatigue, etc. (Farjadian et al. 2019, Zhao et al. 2018). The micellar lotion, upon application, diffuses into blood circulation and exhibits its action faster than other known means (Simon and Group 2006).

I. Nanoparticle/Nanomedicine-based Vaccines

Nanoparticle-based vaccine delivery is also garnering the attention of researchers and health professionals. Nanoparticles not only improve the stability but also the functionality of various vaccines. Some common examples of clinically used NP-based vaccines have been discussed in this section.

Approved in 1994, Epaxal® was the first virosome-based liposomal vaccine, clinically used as an immunizing agent against Hepatitis A. Preservatives like aluminum and thiomersal are lacking in Epaxal® vaccine due to its higher tolerability and decreased undesirable effects compared to conventional vaccines. Hence, Epaxal® can be used intradermally or subcutaneously without any local unwanted effect. A single dose produces immunity for a short while, although the immunity lasts for up to 20 years after the administration of a booster dose (Bovier 2008, Riedemann et al. 2004).

Inflexal® V is a virosome-based adjuvant vaccine composed of deactivated hemagglutinin from strains A and B of the Influenza virus, approved in 1997 by the FDA for the prevention of influenza. This vaccine has an improved safety profile and is less vulnerable to allergic reactions due to the presence of a negligible amount of ovalbumin per dose (less than 10 ng) (Mischler and Metcalfe 2002).

Gardasil® is a quadrivalent human papillomavirus (HPV) vaccine, approved in 2006 for the deterrence of HPV (type 6, type 11, type 16 and type 18) infections such as warts, precancerous lesions, and vaginal and cervical cancer (Roldão et al. 2010). Gardasil® has a good delivery system with a high immune response and enhanced hindrance of the infection (Siddiqui and Perry 2006).

Cervarix® is a bivalent human papillomavirus vaccine, approved in 2009 for the treatment of cancer-caused HPV (type 16 and type 18). Both Gardasil® and Cervarix® are VLP-based formulations; however, yeast is used in the production of Gardasil®, while insect cells are used in the production of Cervarix®. Cervarix® vaccines are safer than Gardasil® due to yeast hypersensitivity in some individuals (Monie et al. 2008).

Pfizer-BioNTech® vaccine was produced by BioNTech co-partnered with Pfizer, which got an emergency use permit from FDA in December 2021, with 90% efficacy against the coronavirus. The vaccine is based on mRNA (genetic molecule) responsible for the production of the spike protein which boosts immunity (Abdellatif et al. 2021, Tamerler and Sarikaya 2009). For stability and enhanced effectiveness, the vaccine (generic name tozinameran) was loaded in PEGylated liposomes (PEGLips) or liposomal nanoparticles (LNPs) (Caballero and Quirce 2021, Kim et al. 2015, Zamboni 2008). Lipid nanoparticle (LNP)-formulated mRNA vaccines have shown prophylactic efficacy against several viral targets in preclinical models (Pardi et al. 2017, Pardi et al. 2019, Sahin et al. 2014); in clinical trials, these vaccines were found to be safe and well tolerated in the prevention of infectious diseases (Sahin et al. 2020). The rapid production process of mRNA and LNP formulation made the development and delivery of pandemic vaccines convenient (Sahin et al. 2020).

Moderna® was the 2nd vaccine against COVID-19, approved by the FDA on an emergency basis on December 18, 2020. Moderna formulated mRNA vaccines from mRNA in LNPs or PEGLips, similar to Pfizer-BioNTech. The liposomes enhanced the stability of the mRNA and mitigated its lability (Kim et al. 2015).

Oxford-AstraZeneca® is a double-stranded DNA vaccine, unlike Moderna's and Pfizer-BioNTech's single-stranded RNA vaccines, and got approval from Britain for emergency use against COVID-19. It showed an efficacy of 82.4% in clinical trials after the administration of the 2nd dose after a span of 12 weeks. The Oxford-AstraZeneca team attached a modified version of a chimpanzee adenovirus to the gene encoding the coronavirus S (spike) protein. Adenoviruses cause flu and cold-like symptoms, and easily enter the cells; but they do not replicate. The vaccine lasts for 6 months if stored at 2–8°C temperatures (Abdellatif and Alsowinea 2021, Medicines and Agency 2020a, 2020b).

Sinopharm® vaccine (inactivated coronavirus) was developed by China at the end of 2020, after the COVID-19 outbreak. The vaccine was found to be 79% effective in clinical trials. It was distributed to other countries after approval. The researchers working on this vaccine disabled coronavirus with beta-propiolactone by binding to their DNA, thus stopping the replication of the virus. The proteins therein, including spikes, activate the immune system to produce antibodies against SARS-CoV-2; the antibodies then bind to spike proteins (Abu-Hammad et al. 2021, Al Khames Aga et al. 2021, Doroftei et al. 2021, Drulovic et al. 2021). Sinopharm® is stored at 2–8°C temperatures.

Novavax®, also known as NVX-CoV2373, is a recombinant vaccine. The genetic sequence of coronavirus (COVID-19) S proteins was used for the production of Novavax® vaccines. These vaccines demonstrated 89% efficacy in preclinical studies (Abdellatif and Alsowinea 2021). Like other COVID-19 vaccines, even these are stored at 2–8°C temperatures.

Johnson and Johnson developed another COVID-19 vaccine known as Ad26.COV2 or JNJ-78436735. In clinical trials, this vaccine demonstrated 72% effectiveness even with a single dose. The vaccine is again based on adenoviruses, like Oxford-AstraZeneca's. The genes for coronavirus S protein were incorporated into an adenovirus which facilitated entry into the cell and did not replicate. The Johnson & Johnson team is performing various trials on viral vectors for other illnesses like HIV, Zika, etc. as well (Orders 2021).

J. Nanoparticles/Nanotechnology as an Imaging/Diagnostic Tool

Diagnosis and detection of diseases is very important in healthcare, and different technologies and probes are used for this purpose. Diagnoses need to be quick, accurate and precise for avoiding 'false negative' results. NP-based diagnosis and imaging are relatively new and more practical than conventional methods. NP-based technology is used in anatomical imaging to obtain facts about the overall structure of tissues and organs, in Computed Tomography (CT) scan or Magnetic Resonance Imaging (MRI) for instance. Similarly, positron emission tomography (PET) or single-photon emission CT (SPECT) provide information about biological and cellular processes like metabolism, DNA synthesis, protein expression, etc. The unique characteristics and size of nanoparticles make them advantageous in diagnosis and imaging, with better targeting approaches, high signal to background ratios, high avidity, high tunability and theranostic capabilities. Earlier, several NP formulations were permitted for clinical use (like Resovist, Feridex, sulfur colloids, etc.), but most of them are now being superseded. Some examples of NP-based preparations used for imaging and diagnosis are mentioned in the succeeding paragraphs.

Nanocoll® is a radiopharmaceutical nanocolloidal (particle size < 100 nm) formulation used for sentinel node localization in breast cancer and in melanoma (O'Brien et al. 2006). Nanocoll® has been registered in Europe since 1995 and is marketed as a kit (lyophilized human albumin and stannous chloride dehydrate). Nanocoll® is administered intravenously for radio imaging like inflammation scintigraphy and bone marrow scintigraphy, and subcutaneously for the determination of lymphatic system integrity (Mitterhauser et al. 2003).

Nanocis® is a radiopharmaceutical nano-colloid preparation of rhenium sulfide (average particle size 50–200 nm), registered in Europe since 2000. Subcutaneous injections of this formulation are used for lymphoscintigraphy, while peri-tumoral or subdermal injections are used for gastro-esophageal scintigraphy and gastroduodenal motor activity (Jimenez et al. 2008).

Table 1: Available therapeutic and diagnostic nanomedicines in the market.

Disease/ Diagnosis	Nanomedicine	Drug Formulation	Application
Cancer	Doxil®	PEGylated-Liposomal Doxorubicin	AIDS-related Kaposi's sarcoma, multiple myeloma and ovarian cancer
	Myocet®	non-PEGylated-Liposomal Doxorubicin	metastatic breast cancer
	DaunoXome®	Liposomal Daunorubicin citrate	HIV-linked Kaposi's sarcoma and acute myeloid leukemia
	Abraxane®	Albuminated Paclitaxel	Breast cancer, pancreatic adenocarcinoma, lung cancer
	Onivyde®	Liposomal Irinotecan	Pancreatic adenocarcinoma
	DepoCyt®	Liposomal Cytarabine	Lymphomatous meningitis
	Marqibo®	Liposomal Vincristine sulfate	Philadelphia chromosome-negative lymphoblastic leukemia
	Genexol-PM®	Polymeric micelle Paclitaxel	Breast cancer and lung cancer
	Vyxeos®	Liposomal Cytarabine and Daunorubicin	Acute myeloid leukemia
	Oncaspars®	PEGasparaginase	Acute lymphocytic leukemia
	Nanoxels®	Micellar Paclitaxel	Solid tumors
	Mepacts®	Liposomal Mifamurtide	Osteogenic sarcoma
	NanoTherms®	Iron oxide NPs	Brain tumors
	Apealeas®	Micellar Paclitaxel	Ovarian, peritoneal, and fallopian tube cancer
	Hensifys®	Hafnium oxide NPs	Locally-advanced soft tissue sarcoma
Fungal infections	AmBisomes®	Liposomal Amphotericin B	Systemic fungal infection
	Amphotecs®	AmB cholesteryl sulfate	Leishmaniasis and fungal infection

Table 1 contd. ...

...Table 1 contd.

Disease/ Diagnosis	Nanomedicine	Drug Formulation	Application
Neurological disorders	Avinzas®	Nanocrystal Morphine sulfate	Psychostimulant, pain relief
	Ritalin LAs®	Nanocrystal Methylphenidate hydrochloride	Attention-deficit/ hyperactivity disorder
	Focalin XRs®	D-threo-enantiomer of methylphenidate	Attention-deficit/ hyperactivity disorder
	Invega Sustennas®	Nanocrystal Paliperidone palmitate	Schizophrenia
Ocular disorders	Visudynes®	Liposomal Verteporfin	Photodynamic therapy (PDT)
	Restasiss®	Anionic nanoemulsion cyclosporine	Severe keratitis
	Novasorbs®	Cationic nanoemulsion cyclosporine	Severe keratitis
Cardiovascular diseases	Tricors®	Nanocrystal Fenofibrate	Hypercholesterolemia or mixed dyslipidemia
	Triglides®	Nanocrystal Fenofibrate	Hypercholesterolemia or mixed dyslipidemia
Blood disorders	Ferahemes®	Paramagnetic iron oxide NPs	Anemia correlated to chronic kidney disease
	Ferinjects®	Iron (III)-oxyhydroxide NPs	Iron deficiency anemia
	Injectafers®	Iron (III)-oxyhydroxide NPs	Iron deficiency anemia
	Monofers®	Iron isomaltoside NPs	Iron deficiency anemia
	Ferumoxytols®	Superparamagnetic iron oxide nanoparticles (SPIONs)	CKD-associated anemia
	Mirceras®	PEGylated erythropoietin beta	Anemia associated with kidney diseases
	Neulastas®	PEGfilgrastim	Febrile neutropenia
	Pegasyss®	PEGinterferon alfa-2a	Hepatitis B and C
	PEGintrons®	PEGinterferon alfa-2b	Hepatitis C

Table 1 contd. ...

...Table 1 contd.

Disease/ Diagnosis	Nanomedicine	Drug Formulation	Application
Immune and hormonal disorders	Copaxones®	Glatiramer acetate and amino acids (L-lysine, L-alanine, L-glutamic acid, and L-tyrosine)	Immunomodulatory, multiple sclerosis
	Estrasorbs®	Micellar NPs Estradiol emulsion	Estrogen therapy, night sweats, disturbed sleep patterns, hot flushes, lethargy and fatigue
Vaccines	Epaxals®	Virosome-based liposomal vaccine	Immunizing agent against Hepatitis A
	Inflexals®	Virosome, Hemagglutinin from strains A and B of Influenza virus	Prevention of influenza
	Gardasils®	HPV quadrivalent vaccine	Warts, precancerous lesions, and vaginal and cervical cancer
	Cervarix®	HPV bivalent vaccine	Cancer caused by HPV
	Pfizer-BioNTech®	mRNA	Coronavirus
	Moderna®	mRNA	Coronavirus
	Oxford-AstraZeneca®	Double-stranded DNA	Coronavirus
	Sinopharm®	Inactivated coronavirus	Coronavirus
	Novavax®	Recombinant vaccine	Coronavirus
	Johnson & Johnson	Double-stranded DNA	Coronavirus

Table 1 contd. ...

...Table 1 contd.

Disease/ Diagnosis	Nanomedicine	Drug Formulation	Application
Imaging/ diagnostic tools	Nanocoll®	Radiolabelled nanocolloid (lyophilized human albumin and stannous chloride dehydrate)	Sentinel node localization in breast cancer and in melanoma, inflammation scintigraphy, bone marrow scintigraphy, lymphatic system integrity
	Nanocis®	Radiolabelled nanocolloid (rhenium sulfide)	Lymphoscintigraphy, gastro-esophageal scintigraphy and gastroduodenal motor activity
	Lymphoscint®	Radiolabelled nanocolloid (tin sulfide)	Lymphoscintigraphy, gastro-esophageal scintigraphy and gastroduodenal motor activity

Lymphoscint® is another radiolabeled nanocolloidal preparation containing tin sulfide (particle size 10–50 nm), which has applications similar to those of Nanocis®. Earlier, numerous superparamagnetic iron oxide nanoparticles (SPIONs) were used for diagnostic purposes as an imaging agent in MRI. Ferumoxytol (FDA approved SPION) is currently used for the treatment of CKD-associated anemia (Giammarile et al. 2013).

8. Future Perspective

Nanomedicine is evolving day by day. Currently, more than 50 FDA or EMA-approved nanoparticle-based medicines as well as imaging agents are available for therapeutic and diagnostic purposes (Table 1). Extensive development in nanomedicines has been observed in the past few decades since the first nanodrug (Doxil®) commercialization and with market trends for cancer therapy. This nanodrug yielded admirable therapeutic results and minimized the adverse effects associated with the conventional analogs as well. In the near future, a generous increase in nanomedicines is expected due to the researchers' curiosity and investors' interest. Presently, the majority of the research work being carried out focuses on the development of anticancer nanoformulations. The results and effectiveness of these formulations have compelled researchers to design and develop NP-based preparations for the management of various other diseases. Most of the nanomedicines available in the market are liposomes and nanocrystal-based formulations. The stability and circulating half-life of many NP-based formulations have improved through PEGylation. Numerous NP-based medicines have been approved in the last few years, while several are in the clinical development stage. The next generation of existing nanomedicines will further accelerate the growth of nanoformulations. Recently, Onpattro (a medication for polyneuropathy) got approval from FDA; this is likely to lay the foundation for the evolution of nucleic-acid-based medicines in gene conveyance for several lethal diseases. Although several factors (like safety, cost, scale-up issues, regulatory hurdles, etc.) affecting the growth of nanomedicines need to be focused on and resolved, establishing proper guidelines for NP-based medicines by FDA, and understanding the complexity of NPs, biological interconnections and characterization tools will aid in facing these challenges and promote the translation of nanomedicines to clinics.

References

Abdellatif, A.A. and Alsowinea, A.F. 2021. Approved and marketed nanoparticles for disease targeting and applications in COVID-19. Nanotechnology Reviews 10(1): 1941–1977.

Abdellatif, A.A., Tawfeek, H.M., Abdelfattah, A., Batiha, G.E.-S. and Hetta, H.F. 2021. Recent updates in COVID-19 with emphasis on inhalation therapeutics: Nanostructured and targeting systems. Journal of Drug Delivery Science and Technology 63: 102435.

Abruzzo, A., Cerchiara, T., Luppi, B. and Bigucci, F. 2019. Transdermal delivery of antipsychotics: rationale and current status. CNS Drugs 33(9): 849–865.

Abu-Hammad, O., Alduraidi, H., Abu-Hammad, S., Alnazzawi, A., Babkair, H., Abu-Hammad, A. et al. 2021. Side effects reported by Jordanian healthcare workers who received COVID-19 vaccines. Vaccines 9(6): 577.

Accomasso, L., Cristallini, C. and Giachino, C. 2018. Risk assessment and risk minimization in nanomedicine: A need for predictive, alternative, and 3Rs strategies. Frontiers in Pharmacology 9: 228.

Aderibigbe, B.A. 2017. Metal-based nanoparticles for the treatment of infectious diseases. Molecules 22(8): 1370.

Al Khames Aga, Q.A., Alkhaffaf, W.H., Hatem, T.H., Nassir, K.F., Batineh, Y., Dahham, A.T., et al. 2021. Safety of COVID-19 vaccines. Journal of Medical Virology 93(12): 6588–6594.

Allen, T.M. and Cullis, P.R. 2004. Drug delivery systems: entering the mainstream. Science 303(5665): 1818–1822.

Allen, T.M. and Cullis, P.R. 2013. Liposomal drug delivery systems: From concept to clinical applications. Advanced Drug Delivery Reviews 65(1): 36–48.

Amiot, C.L., Xu, S., Liang, S., Pan, L. and Zhao, J.X. 2008. Near-infrared fluorescent materials for sensing of biological targets. Sensors 8(5): 3082–3105.

Arvizo, R.R., Bhattacharyya, S., Kudgus, R.A., Giri, K., Bhattacharya, R. and Mukherjee, P. 2012. Intrinsic therapeutic applications of noble metal nanoparticles: Past, present and future. Chemical Society Reviews 41(7): 2943–2970.

Bailey, R.E., Smith, A.M. and Nie, S. 2004. Quantum dots in biology and medicine. Physica E: Low-dimensional Systems and Nanostructures 25(1): 1–12.

Barz, M., Luxenhofer, R. and Schillmeier, M. 2015. Quo vadis nanomedicine? Future Medicine 10: 3089–3091.

Bastian, A.R., Nangarlia, A., Bailey, L.D., Holmes, A., Sundaram, R.V.K., Ang, C. et al. 2015. Mechanism of multivalent nanoparticle encounter with HIV-1 for potency enhancement of peptide triazole virus inactivation. Journal of Biological Chemistry 290(1): 529–543.

Behera, S.S., Patra, J.K., Pramanik, K., Panda, N. and Thatoi, H. 2012. Characterization and evaluation of antibacterial activities of chemically synthesized iron oxide nanoparticles.

Bhatia, S. 2016. Nanoparticles types, classification, characterization, fabrication methods and drug delivery applications. *In*: Natural polymer drug delivery systems, Springer, pp. 33–93.

Biederman, J., Quinn, D., Weiss, M., Markabi, S., Weidenman, M., Edson, K. et al. 2003. Efficacy and safety of Ritalin® LA™, a new, once daily, extended-release dosage form of methylphenidate, in children with attention deficit hyperactivity disorder. Pediatric Drugs 5(12): 833–841.

Bobo, D., Robinson, K.J., Islam, J., Thurecht, K.J. and Corrie, S.R. 2016. Nanoparticle-based medicines: A review of FDA-approved materials and clinical trials to date. Pharmaceutical Research 33(10): 2373–2387.

Boman, N.L., Masin, D., Mayer, L.D., Cullis, P.R. and Bally, M.B. 1994. Liposomal vincristine which exhibits increased drug retention and increased circulation longevity cures mice bearing P388 tumors. Cancer Research 54(11): 2830–2833.

Bovier, P.A. 2008. Epaxal®: A virosomal vaccine to prevent hepatitis A infection. Expert Review of Vaccines 7(8): 1141–1150.

Brintha, S. and Ajitha, M. 2016. Synthesis, structural and antibacterial activity of aluminium and nickel doped ZnO nanoparticles by Sol-gel method. Synthesis 1(1).

Bullivant, J.P., Zhao, S., Willenberg, B.J., Kozissnik, B., Batich, C.D. and Dobson, J. 2013. Materials characterization of Feraheme/ferumoxytol and preliminary evaluation of its potential for magnetic fluid hyperthermia. International Journal of Molecular Sciences 14(9): 17501–17510.

Caballero, M. and Quirce, S. 2021. Excipients as potential agents of anaphylaxis in vaccines: Analyzing the formulations of currently authorized COVID-19 vaccines. J. Investig. Allergol. Clin. Immunol. 31(1): 92–93.

Cai, Z., Wang, Y., Zhu, L.-J. and Liu, Z.-Q. 2010. Nanocarriers: A general strategy for enhancement of oral bioavailability of poorly absorbed or pre-systemically metabolized drugs. Current Drug Metabolism 11(2): 197–207.

Caldwell, J.R. 2004. Avinza®–24-h sustained-release oral morphine therapy. Expert Opinion on Pharmacotherapy 5(2): 469–472.

Casadevall, A. 1996. Antibody-based therapies for emerging infectious diseases. Emerging Infectious Diseases 2(3): 200.

Caster, J.M., Patel, A.N., Zhang, T. and Wang, A. 2017. Investigational nanomedicines in 2016: A review of nanotherapeutics currently undergoing clinical trials. Wiley Interdisciplinary Reviews: Nanomedicine and Nanobiotechnology 9(1): e1416.

Cavalli, R., Gasco, M.R., Chetoni, P., Burgalassi, S. and Saettone, M.F. 2002. Solid lipid nanoparticles (SLN) as ocular delivery system for tobramycin. International Journal of Pharmaceutics 238(1-2): 241–245.

Choi, S.-r., Britigan, B.E. and Narayanasamy, P. 2017. Ga (III) nanoparticles inhibit growth of both Mycobacterium tuberculosis and HIV and release of interleukin-6 (IL-6) and IL-8 in coinfected macrophages. Antimicrobial Agents and Chemotherapy 61(4): e02505–16.

Clemons, K.V. and Stevens, D.A. 1998. Comparison of Fungizone, Amphotec, AmBisome, and Abelcet for treatment of systemic murine cryptococcosis. Antimicrobial Agents and Chemotherapy 42(4): 899–902.

Coimbra, M., Banciu, M., Fens, M.H., de Smet, L., Cabaj, M., Metselaar, J.M. et al. 2010. Liposomal pravastatin inhibits tumor growth by targeting cancer-related inflammation. Journal of Controlled Release 148(3): 303–310.

Crain, M.L. 2018. Daunorubicin and Cytarabine liposome (vyxeos™). Oncology Times 40(10): 30.

Crielaard, B.J., Lammers, T., Schiffelers, R.M. and Storm, G. 2012. Drug targeting systems for inflammatory disease: One for all, all for one. Journal of Controlled Release 161(2): 225–234.

Cuenca, A.G., Jiang, H., Hochwald, S.N., Delano, M., Cance, W.G. and Grobmyer, S.R. 2006. Emerging implications of nanotechnology on cancer diagnostics and therapeutics. Cancer 107(3): 459–466.

Curran, M.P. and McCormack, P.L. 2008. Methoxy polyethylene glycol-epoetin beta. Drugs 68(8): 1139–1156.

Danhier, F. 2016. To exploit the tumor microenvironment: Since the EPR effect fails in the clinic, what is the future of nanomedicine? Journal of Controlled Release 244: 108–121.

Dawidczyk, C.M., Kim, C., Park, J.H., Russell, L.M., Lee, K.H., Pomper, M.G. et al. 2014. State-of-the-art in design rules for drug delivery platforms: Lessons learned from FDA-approved nanomedicines. Journal of Controlled Release 187: 133–144.

Deamer, D.W. 2010. From "banghasomes" to liposomes: A memoir of Alec Bangham, 1921–2010. The FASEB Journal 24(5): 1308.

Doroftei, B., Ciobica, A., Ilie, O.-D., Maftei, R. and Ilea, C. 2021. Mini-review discussing the reliability and efficiency of COVID-19 vaccines. Diagnostics 11(4): 579.

Drulovic, J., Ivanovic, J., Martinovic, V., Tamas, O., Veselinovic, N., Cujic, D. et al. 2021. Humoral response to SARS-CoV-2 COVID-19 vaccines in patients with multiple sclerosis treated with immune reconstitution therapies. Multiple Sclerosis and Related Disorders 54: 103150.

Duggan, S.T. and Keating, G.M. 2011. Pegylated liposomal doxorubicin. Drugs 71(18): 2531–2558.

Dusinska, M., Boland, S., Saunders, M., Juillerat-Jeanneret, L., Tran, L., Pojana, G. et al. 2015. Towards an alternative testing strategy for nanomaterials used in nanomedicine: Lessons from NanoTEST. Nanotoxicology 9(sup1): 118–132.

Farjadian, F., Ghasemi, A., Gohari, O., Roointan, A., Karimi, M. and Hamblin, M.R. 2019. Nanopharmaceuticals and nanomedicines currently on the market: Challenges and opportunities. Nanomedicine 14(1): 93–126.

Fassas, A. and Anagnostopoulos, A. 2005. The use of liposomal daunorubicin (DaunoXome) in acute myeloid leukemia. Leukemia & Lymphoma 46(6): 795–802.

Feng, F., Sakoda, Y., Ohyanagi, T., Nagahori, N., Shibuya, H., Okamastu, M. et al. 2013. Novel thiosialosides tethered to metal nanoparticles as potent influenza A virus haemagglutinin blockers. Antiviral Chemistry and Chemotherapy 23(2): 59–65.

Fütterer, S., Andrusenko, I., Kolb, U., Hofmeister, W. and Langguth, P. 2013. Structural characterization of iron oxide/hydroxide nanoparticles in nine different parenteral drugs for the treatment of iron deficiency anaemia by electron diffraction (ED) and X-ray powder diffraction (XRPD). Journal of Pharmaceutical and Biomedical Analysis 86: 151–160.

Gabizon, A.A., Shmeeda, H. and Zalipsky, S. 2006. Pros and cons of the liposome platform in cancer drug targeting. Journal of Liposome Research 16(3): 175–183.

Gadekar, V., Borade, Y., Kannaujia, S., Rajpoot, K., Anup, N., Tambe, V. et al. 2021. Nanomedicines accessible in the market for clinical interventions. Journal of Controlled Release 330: 372–397.

Gao, L., Liu, G., Ma, J., Wang, X., Zhou, L., Li, X. et al. 2013. Application of drug nanocrystal technologies on oral drug delivery of poorly soluble drugs. Pharmaceutical Research 30(2): 307–324.

Gao, Z., Kennedy, A.M., Christensen, D.A. and Rapoport, N.Y. 2008. Drug-loaded nano/microbubbles for combining ultrasonography and targeted chemotherapy. Ultrasonics 48(4): 260–270.

Gaspar, R. 2007. Regulatory issues surrounding nanomedicines: Setting the scene for the next generation of nanopharmaceuticals.

Gaspar, R. and Duncan, R. 2009. Polymeric carriers: Preclinical safety and the regulatory implications for design and development of polymer therapeutics. Advanced Drug Delivery Reviews 61(13): 1220–1231.

Giammarile, F., Alazraki, N., Aarsvold, J.N., Audisio, R.A., Glass, E., Grant, S.F. et al. 2013. The EANM and SNMMI practice guideline for lymphoscintigraphy and sentinel node localization in breast cancer. European Journal of Nuclear Medicine and Molecular Imaging 40(12): 1932–1947.

Gradishar, W.J., Tjulandin, S., Davidson, N., Shaw, H., Desai, N., Bhar, P. et al. 2005. Phase III trial of nanoparticle albumin-bound paclitaxel compared with polyethylated castor oil–based paclitaxel in women with breast cancer. Journal of Clinical Oncology 23(31): 7794–7803.

Grainger, D.W. 2013. Connecting drug delivery reality to smart materials design. International Journal of Pharmaceutics 454(1): 521–524.

Gregoriadis, G. and Ryman, B. 1971. Liposomes as carriers of enzymes or drugs: A new approach to the treatment of storage diseases. Biochemical Journal 124(5): 58P.

Gupta, A.K. and Gupta, M. 2005. Synthesis and surface engineering of iron oxide nanoparticles for biomedical applications. Biomaterials 26(18): 3995–4021.

Hafner, A., Lovrić, J., Lakoš, G.P. and Pepić, I. 2014. Nanotherapeutics in the EU: An overview on current state and future directions. International Journal of Nanomedicine 9: 1005.

Halder, A., Das, S., Bera, T. and Mukherjee, A. 2017. Rapid synthesis for monodispersed gold nanoparticles in kaempferol and anti-leishmanial efficacy against wild and drug resistant strains. Rsc Advances 7(23): 14159–14167.

Halwani, A.A. 2022. Development of pharmaceutical nanomedicines: From the bench to the market. Pharmaceutics 14(1): 106.

Hang, X., Peng, H., Song, H., Qi, Z., Miao, X. and Xu, W. 2015. Antiviral activity of cuprous oxide nanoparticles against Hepatitis C virus *in vitro*. Journal of Virological Methods 222: 150–157.

Hare, J.I., Lammers, T., Ashford, M.B., Puri, S., Storm, G. and Barry, S.T. 2017. Challenges and strategies in anti-cancer nanomedicine development: An industry perspective. Advanced Drug Delivery Reviews 108: 25–38.

Hassan, N., Boville, B., Reischmann, D., Ndika, A., Sterken, D. and Kovey, K. 2017. Intravenous ferumoxytol in pediatric patients with iron deficiency anemia. Annals of Pharmacotherapy 51(7): 548–554.

Hsueh, Y.-H., Ke, W.-J., Hsieh, C.-T., Lin, K.-S., Tzou, D.-Y. and Chiang, C.-L. 2015. ZnO nanoparticles affect Bacillus subtilis cell growth and biofilm formation. PloS One 10(6): e0128457.

Hua, S. 2013. Targeting sites of inflammation: Intercellular adhesion molecule-1 as a target for novel inflammatory therapies. Frontiers in Pharmacology 4: 127.

Hua, S., De Matos, M.B., Metselaar, J.M. and Storm, G. 2018. Current trends and challenges in the clinical translation of nanoparticulate nanomedicines: Pathways for translational development and commercialization. Frontiers in Pharmacology 9: 790.

Huang, D. and Wu, D. 2018. Biodegradable dendrimers for drug delivery. Materials Science and Engineering: C 90: 713–727.

Iga, A.M., Robertson, J.H., Winslet, M.C. and Seifalian, A.M. 2007. Clinical potential of quantum dots. Journal of Biomedicine and Biotechnology 2007.

Iijima, S. 1991. Helical microtubules of graphitic carbon. Nature 354(6348): 56–58.

Ismail, R.A., Sulaiman, G.M., Abdulrahman, S.A. and Marzoog, T.R. 2015. Antibacterial activity of magnetic iron oxide nanoparticles synthesized by laser ablation in liquid. Materials Science and Engineering: C 53: 286–297.

Jaafar-Maalej, C., Elaissari, A. and Fessi, H. 2012. Lipid-based carriers: Manufacturing and applications for pulmonary route. Expert Opinion on Drug Delivery 9(9): 1111–1127.

Jackman, J.A., Mészáros, T., Fülöp, T., Urbanics, R., Szebeni, J. and Cho, N.-J. 2016. Comparison of complement activation-related pseudoallergy in miniature and domestic pigs: Foundation of a validatable immune toxicity model. Nanomedicine: Nanotechnology, Biology and Medicine 12(4): 933–943.

Jacob Inbaneson, S. and Ravikumar, S. 2013. *In vitro* antiplasmodial activity of PDDS-coated metal oxide nanoparticles against Plasmodium falciparum. Applied Nanoscience 3(3): 197–201.

Jahn, M.R., Andreasen, H.B., Fütterer, S., Nawroth, T., Schünemann, V., Kolb, U. et al. 2011. A comparative study of the physicochemical properties of iron isomaltoside 1000 (Monofer®), a new intravenous iron preparation and its clinical implications. European Journal of Pharmaceutics and Biopharmaceutics 78(3): 480–491.

Jang, B., Kwon, H., Katila, P., Lee, S.J. and Lee, H. 2016. Dual delivery of biological therapeutics for multimodal and synergistic cancer therapies. Advanced Drug Delivery Reviews 98: 113–133.

Jarvis, M., Krishnan, V. and Mitragotri, S. 2019. Nanocrystals: A perspective on translational research and clinical studies. Bioengineering & Translational Medicine 4(1): 5–16.

Jimenez, I.R., Roca, M., Vega, E., García, M.L., Benitez, A., Bajén, M. et al. 2008. Particle sizes of colloids to be used in sentinel lymph node radiolocalization. Nuclear Medicine Communications 29(2): 166–172.

Joseph, E. and Singhvi, G. 2019. Multifunctional nanocrystals for cancer therapy: A potential nanocarrier. Nanomaterials for Drug Delivery and Therapy, 91–116.

Kalele, S., Gosavi, S., Urban, J. and Kulkarni, S. 2006. Nanoshell particles: Synthesis, properties and applications. Current Science, 1038–1052.

Kalra, P.A. 2011. Introducing iron isomaltoside 1000 (Monofer®)—development rationale and clinical experience. NDT Plus 4(suppl_1): i10–i13.

Kalra, P.A. and Bhandari, S. 2016. Efficacy and safety of iron isomaltoside (Monofer®) in the management of patients with iron deficiency anemia. International Journal of Nephrology and Renovascular Disease 9: 53.

Karlsson, J., Vaughan, H.J. and Green, J.J. 2018. Biodegradable polymeric nanoparticles for therapeutic cancer treatments. Annual Review of Chemical and Biomolecular Engineering 9: 105.

Karthik, L., Kumar, G., Keswani, T., Bhattacharyya, A., Reddy, B.P. and Rao, K.B. 2013. Marine actinobacterial mediated gold nanoparticles synthesis and their antimalarial activity. Nanomedicine: Nanotechnology, Biology and Medicine 9(7): 951–960.

Kayser, O., Lemke, A. and Hernandez-Trejo, N. 2005. The impact of nanobiotechnology on the development of new drug delivery systems. Current Pharmaceutical Biotechnology 6(1): 3–5.

Kazi, K.M., Mandal, A.S., Biswas, N., Guha, A., Chatterjee, S., Behera, M. et al. 2010. Niosome: A future of targeted drug delivery systems. Journal of Advanced Pharmaceutical Technology & Research 1(4): 374.

Kesarkar, R., Shroff, S., Yeole, M. and Chowdhary, A. 2015. L-cysteine functionalized gold nanocargos potentiates anti-HIV activity of azidothymydine against HIV-1Ba-L virus. J. Immunol. Virol. 1.

Kherlopian, A.R., Song, T., Duan, Q., Neimark, M.A., Po, M.J., Gohagan, J.K. et al. 2008. A review of imaging techniques for systems biology. BMC Systems Biology 2(1): 1–18.

Kim, H., Park, Y. and Lee, J.B. 2015. Self-assembled messenger RNA nanoparticles (mRNA-NPs) for efficient gene expression. Scientific Reports 5(1): 1–9.

Kim, M. and Williams, S. 2018. Daunorubicin and cytarabine liposome in newly diagnosed therapy-related Acute Myeloid Leukemia (AML) or AML with myelodysplasia-related changes. Annals of Pharmacotherapy 52(8): 792–800.

Kim, S., Chatelut, E., Kim, J.C., Howell, S.B., Cates, C., Kormanik, P.A. et al. 1993. Extended CSF cytarabine exposure following intrathecal administration of DTC 101. Journal of Clinical Oncology 11(11): 2186–2193.

Klibanov, A.L. 2006. Microbubble contrast agents: Targeted ultrasound imaging and ultrasound-assisted drug-delivery applications. Investigative Radiology 41(3): 354–362.

Kraft, J.C., Freeling, J.P., Wang, Z. and Ho, R.J. 2014. Emerging research and clinical development trends of liposome and lipid nanoparticle drug delivery systems. Journal of Pharmaceutical Sciences 103(1): 29–52.

Kuijpers, S.A., Coimbra, M.J., Storm, G. and Schiffelers, R.M. 2010. Liposomes targeting tumour stromal cells. Molecular Membrane Biology 27(7): 328–340.

Kulnigg, S., Stoinov, S., Simanenkov, V., Dudar, L.V., Karnafel, W., Garcia, L.C. et al. 2008. A novel intravenous iron formulation for treatment of anemia in inflammatory bowel disease: the ferric carboxymaltose (FERINJECT®) randomized controlled trial. Official Journal of the American College of Gastroenterology l ACG 103(5): 1182–1192.

Kumar Teli, M., Mutalik, S. and Rajanikant, G. 2010. Nanotechnology and nanomedicine: Going small means aiming big. Current Pharmaceutical Design 16(16): 1882–1892.

Kunjachan, S., Ehling, J., Storm, G., Kiessling, F. and Lammers, T. 2015. Noninvasive imaging of nanomedicines and nanotheranostics: principles, progress, and prospects. Chemical Reviews 115(19): 10907–10937.

Lallemand, F., Schmitt, M., Bourges, J.-L., Gurny, R., Benita, S. and Garrigue, J.-S. 2017. Cyclosporine A delivery to the eye: A comprehensive review of academic and industrial efforts. European Journal of Pharmaceutics and Biopharmaceutics 117: 14–28.

Lammers, T. 2013. Smart drug delivery systems: Back to the future vs. clinical reality. International Journal of Pharmaceutics 454(1): 527–529.

Laverman, P., Carstens, M.G., Boerman, O.C., Dams, E.T.M., Oyen, W.J., van Rooijen, N. et al. 2001. Factors affecting the accelerated blood clearance of polyethylene glycol-liposomes upon repeated injection. Journal of Pharmacology and Experimental Therapeutics 298(2): 607–612.

Lee, M.-Y., Yang, J.-A., Jung, H.S., Beack, S., Choi, J.E., Hur, W. et al. 2012. Hyaluronic acid–gold nanoparticle/interferon α complex for targeted treatment of hepatitis C virus infection. ACS Nano 6(11): 9522–9531.

Lee, S.-M., Chen, H., Dettmer, C.M., O'Halloran, T.V. and Nguyen, S.T. 2007. Polymer-caged lipsomes: A pH-responsive delivery system with high stability. Journal of the American Chemical Society 129(49): 15096–15097.

Leonard, R., Williams, S., Tulpule, A., Levine, A. and Oliveros, S. 2009. Improving the therapeutic index of anthracycline chemotherapy: Focus on liposomal doxorubicin (Myocet™). The Breast 18(4): 218–224.

López, A.M. 2000. Phase II study of liposomal doxorubicin in platinum-paclitaxel refractory epithelial ovarian cancer. Journal of Clinical Oncology 18(17): 3093–3100.

Lu, L., Sun, R.W.-Y., Chen, R., Hui, C.-K., Ho, C.-M., Luk, J.M. et al. 2008. Silver nanoparticles inhibit hepatitis B virus replication. Antiviral Therapy 13(2): 253–262.

Lu, Y., Qi, J., Dong, X., Zhao, W. and Wu, W. 2017. The *in vivo* fate of nanocrystals. Drug Discovery Today 22(4): 744–750.

Lyass, O., Uziely, B., Ben-Yosef, R., Tzemach, D., Heshing, N.I., Lotem, M. et al. 2000. Correlation of toxicity with pharmacokinetics of pegylated liposomal doxorubicin (Doxil) in metastatic breast carcinoma. Cancer: Interdisciplinary International Journal of the American Cancer Society 89(5): 1037–1047.

Lyseng-Williamson, K.A. and Keating, G.M. 2002. Extended-release methylphenidate (Ritalin® LA). Drugs 62(15): 2251–2259.

Maeda, H. 2015. Toward a full understanding of the EPR effect in primary and metastatic tumors as well as issues related to its heterogeneity. Advanced Drug Delivery Reviews 91: 3–6.

Maeda, H., Nakamura, H. and Fang, J. 2013. The EPR effect for macromolecular drug delivery to solid tumors: Improvement of tumor uptake, lowering of systemic toxicity, and distinct tumor imaging *in vivo*. Advanced Drug Delivery Reviews 65(1): 71–79.

Maiseyeu, A., Mihai, G., Kampfrath, T., Simonetti, O.P., Sen, C.K., Roy, S. et al. 2009. Gadolinium-containing phosphatidylserine liposomes for molecular imaging of atherosclerosis. Journal of Lipid Research 50(11): 2157–2163.

Majuru, S. and Oyewumi, M.O. 2009. Nanotechnology in drug development and life cycle management. *In*: Nanotechnology in Drug Delivery, Springer, pp. 597–619.

Malik, T., Chauhan, G., Rath, G., Kesarkar, R.N., Chowdhary, A.S. and Goyal, A.K. 2018. Efaverinz and nano-gold-loaded mannosylated niosomes: A host cell-targeted topical HIV-1 prophylaxis via thermogel system. Artificial Cells, Nanomedicine, and Biotechnology 46(sup1): 79–90.

Mayer, L. Significantly improves overall survival in Phase 3 high-risk AML trial, validating the CombiPlex technology and opening opportunities for novel combinations Lawrence Mayer, Barry Liboiron, Sherwin Xie and Paul Tardi, Kim Paulsen, Michael Chiarella and Arthur Louie. Barry Liboiron, Sherwin Xie and Paul Tardi, Kim Paulsen, Michael Chiarella and Arthur Louie www. controlledreleasesociety. org/meetings/ Documents/2016% 20Abstracts/33. pdf.

Mazayen, Z.M., Ghoneim, A.M., Elbatanony, R.S., Basalious, E.B. and Bendas, E.R. 2022. Pharmaceutical nanotechnology: From the bench to the market. Future Journal of Pharmaceutical Sciences 8(1): 1–11.

McGahan, L. 2008. Continuous erythropoietin receptor activator (Mircera) for renal anemia. Issues in Emerging Health Technologies (113): 1–6.

Medicines, Agency, H.P.R. 2020a. Regulatory approval of COVID-19 Vaccine AstraZeneca.

Medicines, Agency, H.p.R. 2020b. Regulatory approval of Pfizer/BioNTech vaccine for COVID-19.

Meel, R.v.d., Vehmeijer, L.J., Kok, R.J., Storm, G. and van Gaal, E.V. 2016. Ligand-targeted particulate nanomedicines undergoing clinical evaluation: Current status. Intracellular Delivery III, 163–200.

Melamed-Gal, S., Loupe, P., Timan, B., Weinstein, V., Kolitz, S., Zhang, J. et al. 2018. Physicochemical, biological, functional and toxicological characterization of the European follow-on glatiramer acetate product as compared with Copaxone. Eneurologicalsci. 12: 19–30.

Metselaar, J.M., Wauben, M.H., Wagenaar-Hilbers, J.P., Boerman, O.C. and Storm, G. 2003. Complete remission of experimental arthritis by joint targeting of glucocorticoids with long-circulating liposomes. Arthritis & Rheumatism: Official Journal of the American College of Rheumatology 48(7): 2059–2066.

Min, Y., Caster, J.M., Eblan, M.J. and Wang, A.Z. 2015. Clinical translation of nanomedicine. Chemical Reviews 115(19): 11147–11190.

Mischler, R. and Metcalfe, I.C. 2002. Inflexal® V a trivalent virosome subunit influenza vaccine: production. Vaccine 20: B17–B23.

Mishra, Y.K., Adelung, R., Röhl, C., Shukla, D., Spors, F. and Tiwari, V. 2011. Virostatic potential of micro–nano filopodia-like ZnO structures against herpes simplex virus-1. Antiviral Research 92(2): 305–312.

Mitchell, M.J., Billingsley, M.M., Haley, R.M., Wechsler, M.E., Peppas, N.A. and Langer, R. 2021. Engineering precision nanoparticles for drug delivery. Nature Reviews Drug Discovery 20(2): 101–124.

Mitterhauser, M., Wadsak, W., Mien, L.-K., Eidherr, H., Roka, S., Zettinig, G. et al. 2003. The labelling of Nanocoll® with [111In] for dual-isotope scanning. Applied Radiation and Isotopes 59(5-6): 337–342.

Moghassemi, S. and Hadjizadeh, A. 2014. Nano-niosomes as nanoscale drug delivery systems: An illustrated review. Journal of Controlled Release 185: 22–36.

Moghimi, S.M., Hunter, A.C. and Murray, J.C. 2005. Nanomedicine: Current status and future prospects. The FASEB Journal 19(3): 311–330.

Moghimi, S.M. and Hunter, C. 2001. Capture of stealth nanoparticles by the body's defences. Critical Reviews™ in Therapeutic Drug Carrier Systems 18(6).

Mohamed, M.M., Fouad, S.A., Elshoky, H.A., Mohammed, G.M. and Salaheldin, T.A. 2017. Antibacterial effect of gold nanoparticles against Corynebacterium pseudotuberculosis. International Journal of Veterinary Science and Medicine 5(1): 23–29.

Molineux, G. 2004. The design and development of pegfilgrastim (PEG-rmetHuG-CSF, Neulasta®). Current Pharmaceutical Design 10(11): 1235–1244.

Monie, A., Hung, C.-F., Roden, R. and Wu, T.C. 2008. Cervarix™: A vaccine for the prevention of HPV 16, 18-associated cervical cancer. Biologics: Targets and Therapy 2(1): 107.

Moore, R.A., Gaskell, H., Rose, P. and Allan, J. 2011. Meta-analysis of efficacy and safety of intravenous ferric carboxymaltose (Ferinject) from clinical trial reports and published trial data. BMC Blood Disorders 11(1): 1–13.

Mora-Huertas, C.E., Fessi, H. and Elaissari, A. 2010. Polymer-based nanocapsules for drug delivery. International Journal of Pharmaceutics 385(1-2): 113–142.

Morones, J.R., Elechiguerra, J.L., Camacho, A., Holt, K., Kouri, J.B., Ramírez, J.T. et al. 2005. The bactericidal effect of silver nanoparticles. Nanotechnology 16(10): 2346.

Murday, J.S., Siegel, R.W., Stein, J. and Wright, J.F. 2009. Translational nanomedicine: Status assessment and opportunities. Nanomedicine: Nanotechnology, Biology and Medicine, 5(3): 251–273.

Narang, A., Chang, R.-K. and Hussain, M.A. 2013. Pharmaceutical development and regulatory considerations for nanoparticles and nanoparticulate drug delivery systems. Journal of Pharmaceutical Sciences 102(11): 3867–3882.

Narasimha, G., Sridevi, A., Devi Prasad, B. and Praveen Kumar, B. 2014. Chemical synthesis of zinc oxide (ZnO) nanoparticles and their antibacterial activity against a clinical isolate Staphylococcus aureus. International Journal of Nano Dimension 5(4): 337–340.

Nayak, K. and Misra, M. 2018. A review on recent drug delivery systems for posterior segment of eye. Biomedicine and Pharmacotherapy 107: 1564–1582.

Needham, D., Anyarambhatla, G., Kong, G. and Dewhirst, M.W. 2000. A new temperature-sensitive liposome for use with mild hyperthermia: Characterization and testing in a human tumor xenograft model. Cancer Research 60(5): 1197–1201.

Nekkanti, V. and Kalepu, S. 2015. Recent advances in liposomal drug delivery: A review. Pharmaceutical Nanotechnology 3(1): 35–55.

Nel, A., Xia, T., Meng, H., Wang, X., Lin, S., Ji, Z. et al. 2013. Nanomaterial toxicity testing in the 21st century: Use of a predictive toxicological approach and high-throughput screening. Accounts of Chemical Research 46(3): 607–621.

Nichols, J.W. and Bae, Y.H. 2014. EPR: Evidence and fallacy. Journal of Controlled Release 190: 451–464.

Nyström, A.M. and Fadeel, B. 2012. Safety assessment of nanomaterials: Implications for nanomedicine. Journal of Controlled Release 161(2): 403–408.

O'Brien, L.M., Duffin, R. and Millar, A.M. 2006. Preparation of 99mTc-Nanocoll for use in sentinel node localization: Validation of a protocol for supplying in unit-dose syringes. Nuclear Medicine Communications 27(12): 999–1003.

Ojha, T., Pathak, V., Shi, Y., Hennink, W.E., Moonen, C.T., Storm, G. et al. 2017. Pharmacological and physical vessel modulation strategies to improve EPR-mediated drug targeting to tumors. Advanced Drug Delivery Reviews 119: 44–60.

Oomen, A.G., Bos, P.M., Fernandes, T.F., Hund-Rinke, K., Boraschi, D., Byrne, H.J. et al. 2014. Concern-driven integrated approaches to nanomaterial testing and assessment–report of the NanoSafety Cluster Working Group 10. Nanotoxicology 8(3): 334–348.

Orders, M. 2021. FDA Authorizes Johnson & Johnson COVID-19 Vaccine. Med. Lett. Drugs Ther. 63(1620): 41–2.

Oude Blenke, E., Mastrobattista, E. and Schiffelers, R.M. 2013. Strategies for triggered drug release from tumor targeted liposomes. Expert Opinion on Drug Delivery 10(10): 1399–1410.

Pai, A.B. 2017. Complexity of intravenous iron nanoparticle formulations: Implications for bioequivalence evaluation. Annals of the New York Academy of Sciences 1407(1): 17–25.

Pardi, N., Hogan, M.J., Pelc, R.S., Muramatsu, H., Andersen, H., DeMaso, C.R. et al. 2017. Zika virus protection by a single low-dose nucleoside-modified mRNA vaccination. Nature 543(7644): 248–251.

Pardi, N., LaBranche, C.C., Ferrari, G., Cain, D.W., Tombácz, I., Parks, R.J. et al. 2019. Characterization of HIV-1 nucleoside-modified mRNA vaccines in rabbits and rhesus macaques. Molecular Therapy-Nucleic Acids 15: 36–47.

Park, E.J., Amatya, S., Kim, M.S., Park, J.H., Seol, E., Lee, H. et al. 2013. Long-acting injectable formulations of antipsychotic drugs for the treatment of schizophrenia. Archives of Pharmacal. Research 36(6): 651–659.

Park, K. 2017. The drug delivery field at the inflection point: Time to fight its way out of the egg. Journal of Controlled Release 267: 2–14.

Pedziwiatr-Werbicka, E., Serchenya, T., Shcharbin, D., Terekhova, M., Prokhira, E., Dzmitruk, V. et al. 2018. Dendronization of gold nanoparticles decreases their effect on human alpha-1-microglobulin. International Journal of Biological Macromolecules 108: 936–941.

Piedmonte, D.M. and Treuheit, M.J. 2008. Formulation of Neulasta®(pegfilgrastim). Advanced Drug Delivery Reviews 60(1): 50–58.

Prabhu, S. and Poulose, E.K. 2012. Silver nanoparticles: Mechanism of antimicrobial action, synthesis, medical applications, and toxicity effects. International Nano Letters 2(1): 1–10.

Prashanth, P., Raveendra, R., Hari Krishna, R., Ananda, S., Bhagya, N., Nagabhushana, B. et al. 2015. Synthesis, characterizations, antibacterial and photoluminescence studies of solution combustion-derived α-Al2O3 nanoparticles. Journal of Asian Ceramic Societies 3(3): 345–351.

Rafi, M.M., Ahmed, K., Nazeer, K.P., Siva Kumar, D. and Thamilselvan, M. 2015. Synthesis, characterization and magnetic properties of hematite (α-Fe2O3) nanoparticles on polysaccharide templates and their antibacterial activity. Applied Nanoscience 5(4): 515–520.

Raza, M.A., Kanwal, Z., Rauf, A., Sabri, A.N., Riaz, S. and Naseem, S. 2016. Size-and shape-dependent antibacterial studies of silver nanoparticles synthesized by wet chemical routes. Nanomaterials 6(4): 74.

Reddy, K.R., Modi, M.W. and Pedder, S. 2002. Use of peginterferon alfa-2a (40 KD)(Pegasys®) for the treatment of hepatitis C. Advanced Drug Delivery Reviews 54(4): 571–586.

Reddy, L.S., Nisha, M.M., Joice, M. and Shilpa, P. 2014. Antimicrobial activity of zinc oxide (ZnO) nanoparticle against Klebsiella pneumoniae. Pharmaceutical Biology 52(11): 1388–1397.

Reinisch, W., Staun, M., Tandon, R.K., Altorjay, I., Thillainayagam, A.V., Gratzer, C. et al. 2013. A randomized, open-label, non-inferiority study of intravenous iron isomaltoside 1,000 (Monofer) compared with oral iron for treatment of anemia in IBD (PROCEED). The American Journal of Gastroenterology 108(12): 1877.

Reynaldo, G., Rodríguez, L., Menéndez, R., Solazábal, J., Amaro, D., Becquer, M.d.l.A. et al. 2018. A comparative pharmacokinetic and pharmacodynamic study of two novel Cuban PEGylated rHuEPO versus MIRCERA® and ior® EPOCIM. Journal of Pharmacy and Pharmacognosy Research 6(3): 179.

Riedemann, S., Reinhardt, G., Ibarra, H. and Frösner, G. 2004. Immunogenicity and safety of a virosomal hepatitis A vaccine (Epaxal®) in healthy toddlers and children in Chile. Acta Paediatrica 93(3): 412–414.

Rizzo, L.Y., Theek, B., Storm, G., Kiessling, F. and Lammers, T. 2013. Recent progress in nanomedicine: Therapeutic, diagnostic and theranostic applications. Current Opinion in Biotechnology 24(6): 1159–1166.

Roldão, A., Mellado, M.C.M., Castilho, L.R., Carrondo, M.J. and Alves, P.M. 2010. Virus-like particles in vaccine development. Expert Review of Vaccines 9(10): 1149–1176.

Romaniuk, J.A. and Cegelski, L. 2015. Bacterial cell wall composition and the influence of antibiotics by cell-wall and whole-cell NMR. Philosophical Transactions of the Royal Society B: Biological Sciences 370(1679): 20150024.

Ryoo, S.-R., Jang, H., Kim, K.-S., Lee, B., Kim, K.B., Kim, Y.-K. et al. 2012. Functional delivery of DNAzyme with iron oxide nanoparticles for hepatitis C virus gene knockdown. Biomaterials 33(9): 2754–2761.

Saad, M.Z.H., Jahan, R. and Bagul, U. 2012. Nanopharmaceuticals: A new perspective of drug delivery system. Asian Journal of Biomedical and Pharmaceutical Sciences 2(14): 11.

Sadiq, I.M., Chowdhury, B., Chandrasekaran, N. and Mukherjee, A. 2009. Antimicrobial sensitivity of Escherichia coli to alumina nanoparticles. Nanomedicine: Nanotechnology, Biology and Medicine 5(3): 282–286.

Saha, B., Bhattacharya, J., Mukherjee, A., Ghosh, A., Santra, C., Dasgupta, A.K. et al. 2007. In vitro structural and functional evaluation of gold nanoparticles conjugated antibiotics. Nanoscale Research Letters 2(12): 614–622.

Sahin, U., Karikó, K. and Türeci, Ö. 2014. mRNA-based therapeutics—Developing a new class of drugs. Nature Reviews Drug Discovery 13(10): 759–780.

Sahin, U., Muik, A., Derhovanessian, E., Vogler, I., Kranz, L.M., Vormehr, M. et al. 2020. COVID-19 vaccine BNT162b1 elicits human antibody and TH1 T cell responses. Nature 586(7830): 594–599.

Sahoo, S.K. and Labhasetwar, V. 2003. Nanotech approaches to drug delivery and imaging. Drug Discovery Today 8(24): 1112–1120.

Sainz, V., Conniot, J., Matos, A.I., Peres, C., Zupanðið, E., Moura, L. et al. 2015. Regulatory aspects on nanomedicines. Biochemical and Biophysical Research Communications 468(3): 504–510.

Sarid, R., Gedanken, A. and Baram-Pinto, D. 2014. Pharmaceutical compositions comprising water-soluble sulfonate-protected nanoparticles and uses thereof, Google Patents.

Sawant, R.R. and Torchilin, V.P. 2012. Challenges in development of targeted liposomal therapeutics. The AAPS Journal 14(2): 303–315.

Saylor, C., Dadachova, E. and Casadevall, A. 2009. Monoclonal antibody-based therapies for microbial diseases. Vaccine 27: G38–G46.

Sercombe, L., Veerati, T., Moheimani, F., Wu, S.Y., Sood, A.K. and Hua, S. 2015. Advances and challenges of liposome assisted drug delivery. Frontiers in Pharmacology 6: 286.

Sessa, G. and Weissmann, G. 1968. Phospholipid spherules (liposomes) as a model for biological membranes. Journal of Lipid Research 9(3): 310–318.

Shah, H.S., Khalid, F., Bashir, S., Asad, M.H.B., Khan, K.-U.-R., Usman, F. et al. 2019. Emulsion-templated synthesis and *in vitro* characterizations of niosomes for improved therapeutic potential of hydrophobic anti-cancer drug: tamoxifen. Journal of Nanoparticle Research 21(2): 1–10.

Shi, J., Kantoff, P.W., Wooster, R. and Farokhzad, O.C. 2017. Cancer nanomedicine: Progress, challenges and opportunities. Nature Reviews Cancer 17(1): 20–37.

Siddiqui, M.A.A. and Perry, C.M. 2006. Human papillomavirus quadrivalent (types 6, 11, 16, 18) recombinant vaccine (Gardasil®). Drugs 66(9): 1263–1271.

Silverman, J.A. and Deitcher, S.R. 2013. Marqibo® (vincristine sulfate liposome injection) improves the pharmacokinetics and pharmacodynamics of vincristine. Cancer Chemotherapy and Pharmacology 71(3): 555–564.

Simon, J.A. and Group, E.S. 2006. Estradiol in micellar nanoparticles: The efficacy and safety of a novel transdermal drug-delivery technology in the management of moderate to severe vasomotor symptoms. Menopause 13(2): 222–231.

Singh, K.K. and Vingkar, S.K. 2008. Formulation, antimalarial activity and biodistribution of oral lipid nanoemulsion of primaquine. International Journal of Pharmaceutics 347(1-2): 136–143.

Smekalova, M., Aragon, V., Panacek, A., Prucek, R., Zboril, R. and Kvitek, L. 2016. Enhanced antibacterial effect of antibiotics in combination with silver nanoparticles against animal pathogens. The Veterinary Journal 209: 174–179.

Strasfeld, L. and Chou, S. 2010. Antiviral drug resistance: Mechanisms and clinical implications. Infectious Disease Clinics 24(3): 809–833.

Svenson, S. 2012. Clinical translation of nanomedicines. Current Opinion in Solid State and Materials Science 16(6): 287–294.

Szebeni, J. 2005. Complement activation-related pseudoallergy: A new class of drug-induced acute immune toxicity. Toxicology 216(2-3): 106–121.

Szebeni, J. and Moghimi, S.M. 2009. Liposome triggering of innate immune responses: A perspective on benefits and adverse reactions: biological recognition and interactions of liposomes. Journal of Liposome Research 19(2): 85–90.

Szebeni, J. and Storm, G. 2015. Complement activation as a bioequivalence issue relevant to the development of generic liposomes and other nanoparticulate drugs. Biochemical and Biophysical Research Communications 468(3): 490–497.

Tamerler, C. and Sarikaya, M. 2009. Molecular biomimetics: Nanotechnology and bionanotechnology using genetically engineered peptides. Philosophical Transactions of the Royal Society A: Mathematical, Physical and Engineering Sciences 367(1894): 1705–1726.

Taniguchi, N. 1974. On the basic concept of nanotechnology. Proceeding of the ICPE.

Theek, B., Gremse, F., Kunjachan, S., Fokong, S., Pola, R., Pechar, M. et al. 2014. Characterizing EPR-mediated passive drug targeting using contrast-enhanced functional ultrasound imaging. Journal of Controlled Release 182: 83–89.

Thomas, D.A., Sarris, A.H., Cortes, J., Faderl, S., O'Brien, S., Giles, F.J. et al. 2006. Phase II study of sphingosomal vincristine in patients with recurrent or refractory adult acute lymphocytic leukemia. Cancer 106(1): 120–127.

Tinkle, S., McNeil, S.E., Mühlebach, S., Bawa, R., Borchard, G., Barenholz, Y. et al. 2014. Nanomedicines: Addressing the scientific and regulatory gap. Annals of the New York Academy of Sciences 1313(1): 35–56.

Torchilin, V. 2004. Targeted polymeric micelles for delivery of poorly soluble drugs. Cellular and Molecular Life Sciences CMLS 61(19): 2549–2559.

Torchilin, V.P. 2006. Multifunctional nanocarriers. Advanced Drug Delivery Reviews 58(14): 1532–1555.

Torchilin, V.P. 2005. Recent advances with liposomes as pharmaceutical carriers. Nature Reviews Drug Discovery 4(2): 145–160.

Trigilio, J., Antoine, T.E., Paulowicz, I., Mishra, Y.K., Adelung, R. and Shukla, D. 2012. Tin oxide nanowires suppress herpes simplex virus-1 entry and cell-to-cell membrane fusion. PLoS One 7(10): e48147.

Voicu, G., Oprea, O., Vasile, B. and Andronescu, E. 2013. Antibacterial activity of zinc oxide-gentamicin hybrid material. Digest Journal of Nanomaterials & Biostructures (DJNB) 8(3).

Voltan, A.R., Quindos, G., Alarcón, K.P.M., Fusco-Almeida, A.M., Mendes-Giannini, M.J.S. and Chorilli, M. 2016. Fungal diseases: Could nanostructured drug delivery systems be a novel paradigm for therapy? International Journal of Nanomedicine 11: 3715.

Vonka, V., Kanaka, J., Hirsch, I., Zavadova, H., Krčmář, M., Suchankova, A. et al. 1984. Prospective study on the relationship between cervical neoplasia and herpes simplex type-2 virus. II. Herpes simplex type-2 antibody presence in sera taken at enrolment. International Journal of Cancer 33(1): 61–66.

Wagner, V., Dullaart, A., Bock, A.-K. and Zweck, A. 2006. The emerging nanomedicine landscape. Nature Biotechnology 24(10): 1211–1217.

Weissig, V., Elbayoumi, T., Flühmann, B. and Barton, A. 2021. The growing field of nanomedicine and its relevance to pharmacy curricula. American Journal of Pharmaceutical Education 85(8).

Weissig, V. and Guzman-Villanueva, D. 2015. Nanopharmaceuticals (part 2): Products in the pipeline. International Journal of Nanomedicine 10: 1245.

Weissig, V., Pettinger, T.K. and Murdock, N. 2014. Nanopharmaceuticals (part 1): products on the market. International Journal of Nanomedicine 9: 4357.

Weng, Y., Liu, J., Jin, S., Guo, W., Liang, X. and Hu, Z. 2017. Nanotechnology-based strategies for treatment of ocular disease. Acta Pharmaceutica Sinica B 7(3): 281–291.

Werner, M.E., Cummings, N.D., Sethi, M., Wang, E.C., Sukumar, R., Moore, D.T. et al. 2013. Preclinical evaluation of Genexol-PM, a nanoparticle formulation of paclitaxel, as a novel radiosensitizer for the treatment of non-small cell lung cancer. International Journal of Radiation Oncology* Biology* Physics 86(3): 463–468.

West, J.L. and Halas, N.J. 2000. Applications of nanotechnology to biotechnology: Commentary. Current Opinion in Biotechnology 11(2): 215–217.

Yang, S.C., Lu, L.F., Cai, Y., Zhu, J.B., Liang, B.W. and Yang, C.Z. 1999. Body distribution in mice of intravenously injected camptothecin solid lipid nanoparticles and targeting effect on brain. Journal of Controlled Release 59(3): 299–307.

Zahir, A.A., Chauhan, I.S., Bagavan, A., Kamaraj, C., Elango, G., Shankar, J. et al. 2015. Green synthesis of silver and titanium dioxide nanoparticles using Euphorbia prostrata extract shows shift from apoptosis to G0/G1 arrest followed by necrotic cell death in Leishmania donovani. Antimicrobial Agents and Chemotherapy 59(8): 4782–4799.

Zain, N.M., Stapley, A.G. and Shama, G. 2014. Green synthesis of silver and copper nanoparticles using ascorbic acid and chitosan for antimicrobial applications. Carbohydrate Polymers 112: 195–202.

Zamboni, W.C. 2008. Concept and clinical evaluation of carrier-mediated anticancer agents. The Oncologist 13(3): 248–260.

Zhang, H. 2016. Onivyde for the therapy of multiple solid tumors. OncoTargets and Therapy 9: 3001.

Zhang, L., Gu, F., Chan, J., Wang, A., Langer, R. and Farokhzad, O. 2008. Nanoparticles in medicine: Therapeutic applications and developments. Clinical Pharmacology & Therapeutics 83(5): 761–769.

Zhao, N., Woodle, M.C. and Mixson, A.J. 2018. Advances in delivery systems for doxorubicin. Journal of Nanomedicine & Nanotechnology 9(5).

Zhou, Y., Kong, Y., Kundu, S., Cirillo, J.D. and Liang, H. 2012. Antibacterial activities of gold and silver nanoparticles against *Escherichia coli* and bacillus Calmette-Guérin. Journal of Nanobiotechnology 10(1): 1–9.

CHAPTER 7

Occupational Exposure to Nanomaterials

Hazards, Toxicity and Safety Concerns

Nafiseh Nasirzadeh and *Soqrat Omari Shekaftik**

1. Introduction

Recently, the development of nanomaterials has increased and presents a wide range of new opportunities in different fields of research and development, which encompass health, the environment and manufacturing technology (Bergamaschi 2009). As a result, the emerging subject of concern on hazards, toxicity and safety of nanomaterials is demonstrated (Llaver et al. 2021).

The major risk to health posed by NMs appears to be related to the exposure to them. For example, occupational exposure to NMs can occur during their industrial or small-scale production and waste discharge (Kuempel et al. 2012). Among exposure scenarios, the workplaces where NMs are produced, processed, used and disposed, are the particular challenges. Some studies have confirmed the potential exposure to NMs is through the inhalation, cutaneous and ocular pathways (Nthwane et al. 2019). Also, injections of NMs may be accidental (e.g., needle stick) or intentional (e.g., drug delivery devices). So, in such a situation, nanomaterials are widely distributed through the body. If they spread into the bloodstream, they may accumulate in different organs (Oberdörster et al. 2005).

Concerns have also increased about NMs released into the environment (Llaver et al. 2021). Investigations showed that NMs can accumulate in

School of Public Health, Department of Occupational Health Engineering, Tehran University of Medical Sciences, Tehran, Iran.
* Corresponding author: S-omarish@razi.tums.ac.ir

the most important natural resources such as air, water and soil (Chen et al. 2011). As a consequence, they can be confined by sediment-dwelling animals or filter feeders and humans can be indirectly influenced by them (Zhang et al. 2022).

Toxicity of NMs is dependent on different factors such as exposure dosage, surface reactivity, surface area, size, shape, and surface charge (Nasirzadeh et al. 2019). Most studies conducted on them are related to *in-vitro* systems and human studies are either non-existent or very few (Omari Shekaftik and Nasirzadeh 2021). However, these studies reported that NMs can lead to the production of active oxidative species, DNA faction, genotoxicity mechanisms, and physical destruction, and epigenetic toxicity, etc. (Gomez-Villalba et al. 2023).

Safety concerns about the NMs is another important issue. Fire, explosion and other unexpected reactions are major safety hazards (Khan 2013). The processes of producing NMs need to the use of highly reactive materials including organic solvents and strong oxidizers (Dreizin 2009). So, these materials can pose a higher risk than bulk materials due to the increased surface-area-to-volume ratio (Eckhoff 2013).

Since they may show characteristics different from their conventional form, there is a need to consider occupational health and safety measures for adequate protection (Savolainen et al. 2010). In this line, exposure control is an important strategy for eliminating or decreasing the exposure. Exposure control consists of a standardized hierarchy, including elimination, substitution, isolation, engineering controls, administrative controls and personal protective equipment (PPE) (Freeland et al. 2016).

Considering the importance of occupational exposure to nanomaterials, hazards, toxicity and safety concerns of them, the aim of this chapter is discusses on these issues.

2. Hazards of Nanomaterials

Our cognition of the occupational and environmental health and safety aspects of nanomaterials is still incomplete and in its formative stage (Omari Shekaftik et al. 2020). However, there is ample evidence of the harmful effects of nanomaterials. The rapid commercialization of nanotechnology necessitates thoughtful and attentive research about the effects of nanomaterials on health and safety (Pattan and Kaul, 2014). This section describes the environmental hazards and safety concerns induced by nanomaterials, in addition to the biological interactions of nanomaterials.

2.1 Interactions with the Human Body

Exposure Scenarios

There should be significant efforts to prevent or minimize exposure to nanomaterials. Factors affecting exposure to nanomaterials include the amount of materials being used and whether they can be easily dispersed or form airborne sprays or droplets. The degree of containment and duration of activity will also influence exposure. Droplet size is another factor affecting the penetration of nanomaterials. At present, there is insufficient information to predict all of the workplace scenarios that are likely to lead to exposure to nanomaterials (Omari Shekaftik et al. 2022a, Omari Shekaftik et al. 2022b). However, there are some workplace factors that increase the potential for exposure to nanomaterials:

- Working with nanomaterials in liquid media without adequate protection (e.g., gloves).
- Working with nanomaterials in liquid media during pouring or mixing operations, or where a high degree of agitation is involved.
- Machining, sanding, drilling or other mechanical disruptions of nanomaterials.
- Generating nanomaterials in the gas phase in non-enclosed systems
- Handling nano-powders (e.g., weighing, blending, spraying).
- Maintenance of equipment and processes used to fabricate or use nanomaterials.
- Cleaning of dust collection systems used to capture nanomaterials.
- Cleaning up of spills and waste materials containing nanomaterials.

Different Physical States of Nanomaterials and their Impact upon Exposure

Nanomaterials can exist as nano-powders, or be suspended in air (ultrafine particles, nanoparticles, aerosols), suspended in liquids (colloids) and incorporated in solids. For biological safety evaluation, manufactured nanomaterials need to be dispersed in an appropriate media (Hodson et al. 2009). The interaction between these media and the nanomaterials can have a profound influence on the behavior of the suspension. Also, the form of the nanomaterial plays a key role in the exposure potential (Christian et al. 2008).

With the increasing number of new manufactured nanomaterials, the importance of dissolution kinetics has risen. Because dissolution kinetics is generally proportional to the surface area, nanomaterials are likely to dissolve much more rapidly than bulk materials (Borm et al. 2006).

Routes of Exposure

Nanomaterials are capable of moving past the protective mechanisms of the body and lead to the body's response or serious adverse health effects (Long et al. 2022). Some exposures occur intentionally (i.e., injection of drug delivery devices, application of skin products, etc.), whereas certain others are unintentional exposures (through inhalation, dermal contact, and ingestion) (Banoun 2022). So, the risk increases with the duration of exposure to nanomaterials and their concentration (Madannejad et al. 2019) (Figure 1).

- Inhalation

Inhalation is the most common route of exposure to nanomaterials in workplaces (Omari Shekaftik et al. 2022b). The deposition of nanoparticles in the respiratory pathway is related to the particle's aerodynamic and thermodynamic diameter– the particle shape and size. Aerodynamic diameter affects agglomeration of particles and the degree of agglomeration can affect the toxicity of the inhaled nanomaterials (Park et al. 2018).

Owing to their small size, nanoparticles can penetrate cell membranes and interfere with cell processes (Oroojalian et al. 2021). Inhalation of

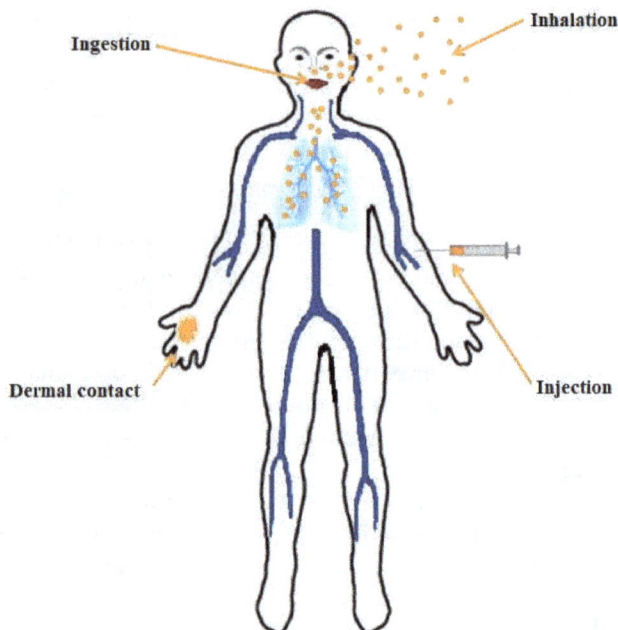

Figure 1: Routes of exposure to nanomaterials.

nanomaterials can result in pulmonary inflammatory reactions in the lungs, which may in turn cause asthma, fibrosis, cancer, or necrosis of lung tissues (Ghumman et al. 2021). This inflammation may also allow nanomaterials to reach nearby lymph nodes, where they may accumulate or translocate to other organs (Seiffert et al. 2022).

According to animal studies, discrete nanoparticles, upon inhalation, may enter the bloodstream from the lungs and translocate to the brain via nasal passages (Alhamoud et al. 2022). Studies have shown that the inhaled nanomaterials can induce certain cancers and cause cardiovascular dysfunction and/or rapid and persistent pulmonary fibrosis (Dong and Ma 2019).

- Dermal Contact

Dermal penetration could occur when the skin is exposed to nanomaterials during the manufacture and use of nanomaterials, or contact with contaminated surfaces (Basinas et al. 2018). Studies show that nanomaterials with different physicochemical properties are able to penetrate the intact skin of pigs (Gimeno-Benito et al. 2021). On the other hand, nanomaterials can penetrate the stratum corneum barrier by passive diffusion and localize within the epidermal and dermal layers within 8–24 hours (Kis et al. 2022).

Some studies on animal models have shown the adverse effects of the penetration of nanomaterials into skin cells. In this domain, studies conducted on *in vitro* models have shown that both single-walled carbon nanotubes (SWCNTs) and multi-walled carbon nanotubes (MWCNT) can enter cells and cause the release of pro-inflammatory cytokines, oxidative stress, and decreased viability (Young et al. 2021). However, it remains unclear as to how these findings may be extrapolated to a potential occupational risk, given that additional data is not yet available for comparing the cell model studies with actual conditions of occupational exposure (Iavicoli et al. 2014).

- Ingestion

Ingestion is another route whereby nanomaterials may enter the body (Yah et al. 2012). Ingestion may occur in conjunction with inhalation or may occur as a result of hand-to-mouth transfer or contamination of food and drinks (Viegas 2020). After ingestion, some nanomaterials are absorbed by the digestive system and transported to lymphatic tissues. Others may accumulate in the digestive system, potentially leading to gastrointestinal blockages (Florence 1997).

This is specifically true while handling porous dry nanomaterials with very high surface areas (> 20 m/g), which includes metals, non-metal oxides, or their mixed forms, and microporous or mesoporous metal organic frameworks (Sharma et al. 2022, Wang et al. 2020).

• Injection

Injections of nanomaterials may be accidental (e.g., needle stick) or intentional (e.g., drug delivery devices). Intravenously injected nanomaterials are widely distributed through the body. If they spread into the bloodstream, they may accumulate in different organs such as the liver, spleen, lungs and kidneys (Oberdörster et al. 2005). In fact, the extent of a nanomaterial's distribution in tissues is greatly modulated by particle and agglomerate size. One study showed that 24 hours after the intravenous injection of gold nanomaterials in mice, at different degrees of agglomeration/aggregation (5–8 nm, 30–200 nm or 500–2,000 nm), tissue distributions were different and the said nanomaterials had accumulated in the liver, spleen and lungs. In another study, rats were injected with gold nanoparticles; the results showed that nanomaterials had the most widespread distribution in the brain, thymus and testes (Bruinink et al. 2015).

2.2 *Environmental Hazards*

Nowadays, concerns have increased about the transport transformation and fate of nanoparticles released into the environment (Figure 2) (Llaver et al. 2021). Generally, sources contributing to the formation of nanoparticles in the environment include: stationary sources (e.g., industrial processes, combustion systems, incinerators), mobile sources (e.g., transportation vehicles), natural sources (e.g., forest fires, volcanic eruptions) and medical sources (e.g., drug delivery via nanoparticles) (De Jong and Borm 2008). In fact, emitted nanoparticles can accumulate in the environment (air, water and soil), and cause environmental and health effects (Chen et al. 2011).

Aggregation and sedimentation of nanomaterials in the environment affects the concentration of free nanoparticles. Aggregated materials are less mobile, but can be trapped by sediment-dwelling animals or filter feeders. So, humans can be indirectly influenced by consuming plants or animals which have accumulated nanoparticles (Zhang et al. 2022).

In the environment, the released nanoparticles are affected by environmental agents such as light, oxidants and microorganisms. This can result in chemical or biological alterations and may lead to the release of nanoparticles (Nowack and Bucheli 2007). The surface of pristine nanoparticles can also be modified by environmental factors such as coating by organic materials or functionalization by chemical or biological processes. For example, the effect of humic and fulvic acids to inhibit the

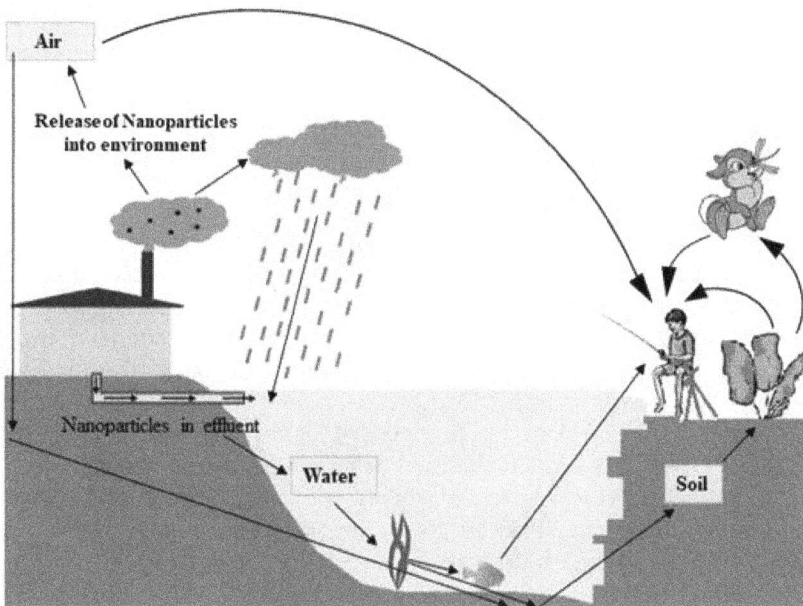

Figure 2: Transport, transformation and fate of nanoparticles released into the environment.

aggregation of carbon nanotubes has been recently demonstrated; also, ZnO nanoparticles can be coated with the surfactant sodium dodecyl sulfate, and can be stable in soil suspension for 14 days, without any change in the particle size distribution (Guldi et al. 1994, Gimbert et al. 2007)

Currently, there is no actual data on the concentration of nanomaterials in the environment, and definitely none on their forms or distribution. Published quantitative studies on the uptake and accumulation of nanomaterials by whole organisms is rare (Klaine et al. 2008). It is clear that organisms living in environments containing nanomaterials can combine them within their bodies (Roberts et al. 2007). So, nanomaterials can enter cells by diffusing via cell membranes as well as through endocytosis and adhesion to cells (Lin et al. 2006). Some nanomaterials, such as quantum dots and carbon nanotubes, are purposely designed to interact with proteins, nucleic acids or cell membranes for labeling or drug delivery (Klaine et al. 2008). Additionally, bacteria can be used to transport nanomaterials. However, the unintentional interactions are more relevant to environmental impacts because they are not controlled and can prove to be detrimental (Diao et al. 2005).

By increasing the control over released nanomaterials in processes such as spillages associated with the transportation of manufactured nanomaterials from production facilities, intentional releases for environmental applications, and releases associated with wear and erosion from general use (the biggest sources for environmental release), it is possible to prevent harmful effects and environmental problems (Klaine et al. 2008).

2.3 Safety Concerns

Fire, explosion and other unexpected reactions involving nanomaterials are major safety hazards (Khan 2013). The processes of producing nanomaterials require the use of a variety of highly reactive materials such as concentrated mineral acids, organic solvents and strong oxidizers (Dreizin 2009). Nanoscale combustible materials can pose a higher risk than bulk materials with a similar mass because of the increased surface-area-to-volume ratio (Eckhoff 2013).

A study undertaken by the HSE (Health and Safety Executive) indicates that nanomaterials can be spontaneously flammable on exposure to air. This is particularly the case with metal nanoparticles as they oxidase easily (Eckhoff 2003). Dusts, like gases and vapors, can form explosive clouds only if the dust concentration lies within certain limits– lower explosion limit (LEL) and upper explosion limit (UEL). For dusts, the LEL is sometimes referred to as the minimum explosive concentration (MEC). The explosion limits, particularly the upper limit, for dusts are not as well defined as for gases and vapors. These limits depend on the dust composition, the particle size distribution and the method of determination. The effect of particle size is likely to be much more marked for nanomaterials. It is anticipated that nanomaterials are more easily ignited than bulk materials (Pritchard 2004). One study indicates that the explosion hazard from aluminum nano-powders is less than that from coarser micron-sized aluminum powders (Li et al. 2016). However, the relationships with particle size are not linear (Pritchard 2004).

Also, nanomaterials may unexpectedly become chemical catalysts and result in unanticipated reactions (Wang et al. 2022). In fact, a catalyst can change the rate of a chemical reaction while not being consumed by the reaction itself; some nanomaterials may initiate catalytic reactions (Xiang et al. 2020). If a material is innately a catalyst, it would be more efficient with small particle sizes, but it is possible that a non-catalyst would become a catalyst by virtue of becoming small/nano-sized. For example, a reduction in particle size of gold nanoparticles can lead to the conversion of a non-catalyst into a catalyst, and provide the potential for unexpected adverse reactions due to catalysis (Neltner 2010).

Since nanomaterials may show characteristics different from their conventional form, there is a need to consider occupational health and safety measures for adequate protection (Savolainen et al. 2010). Monitoring is a key section of the safety program to protect workers against nanomaterial exposures. Monitoring is carried out in personal, regional or biological ways. Personal monitoring is the most common, but regional and biological monitoring also serve important purposes in ensuring occupational health. Personal monitoring is a measure for evaluating the level of exposure to workers, whereas area monitoring is a measure for evaluating the cleanliness level in the work environment (Iavicoli et al. 2018). Moreover, careful management of nanomaterials must be an important consideration using prudent workplace practices. All persons working with nanomaterials must have chemical safety training and/or refresher up-to-date (Council 2011). Fire prevention should also take into account the existing regulations, especially electrical requirements regarding intrinsic safety (Pfaff 2022).

3. Toxicological Effects

Different factors affect the toxicity of nanomaterials, the most important ones being exposure dosage, surface reactivity, surface area, size and shape of nanomaterials, and surface charge (Nasirzadeh et al. 2019). Studies have shown that the toxicity of nanomaterials is mainly applied at the cellular level, and toxicity evaluations are mostly managed in *in vitro* and *in vivo* systems. The two main forms of toxicity at this level are cytotoxicity and genotoxicity (Omari Shekaftik and Nasirzadeh 2021). Figure 3 shows cytotoxicity and genotoxicity mechanisms for nanomaterials.

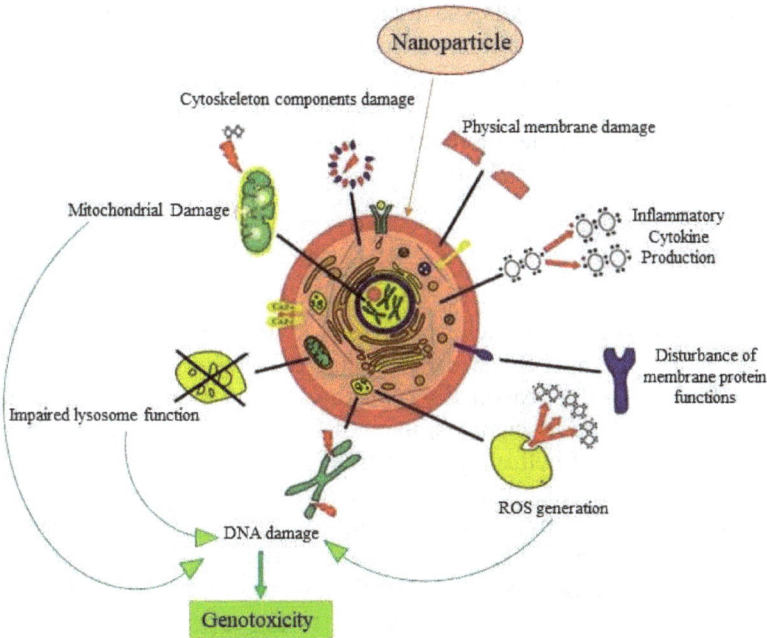

Figure 3: Cytotoxicity and genotoxicity mechanisms for nanomaterials.

3.1 Carbon-based Nanomaterials

All nanomaterials composed of carbon atoms are called carbon-based or carbon nanomaterials. At present, these materials are a novel class of materials that are wildly used in many fields of science and technology, such as medical science, biological science, physical science, chemical science, agriculture, construction, communication, military, etc. (Ghafari et al. 2020). *Fullerene, graphene and carbon nanotubes (CNTs)* are the main nanomaterials derived from carbon. In fullerene, single and double bonds form a closed mesh with fused rings of five to seven atoms. So, a fullerene molecule may be a hollow sphere, ellipsoid, tube or many other shapes, and of different sizes (Malhotra et al. 2020). Graphene is an allotrope of carbon, consisting of a single layer of atoms arranged in a two-dimensional honeycomb lattice nanostructure (Mukhopadhyay et al. 2017). When the basic form of graphene is manipulated and rolled into cylinders, CNTs are formed. The sheets of graphene used to make these nanotubes are 2D due to graphene being one atom thick; this gives the CNTs some of their special properties (Neto et al. 2006). If single-layer graphene is rolled, it can form single-walled CNTs (SWCNTs). Also, rolling several layers of these sheets can create multi-walled carbon nanotubes (MWCNTs), which have slightly different properties (Aqel et al. 2012).

Because of the unique physicochemical properties and the wide applications of carbon-based nanomaterials, exposure to them can occur for a number of individuals (Nasirzadeh et al. 2020). Canu et al. (2016) in a review paper expressed that academic and private laboratory workers, primary and secondary manufacturers, and CNT buyers are facing occupational exposure to carbon nanomaterials (Canu et al. 2016). So, these days, considerable attention is being paid to the toxicity issues of carbon-based nanomaterials. However, there are still uncertainties about their toxicity.

3.1.1 Fullerenes

In 1985, the first fullerene was discovered by Sir Harold W. Kroto of the United Kingdom, and Richard E. Smalley and Robert F. Curl, Jr. of the United States. For this reason, they were awarded the Nobel Prize for their efforts, in 1996. These scientists, using a laser to vaporize graphite rods in an atmosphere of helium gas, obtained cage-like molecules composed of 60 carbon atoms (C60) joined together by single and double bonds to form a hollow sphere with 12 pentagonal and 20 hexagonal faces (an icosahedron)– a design that was similar to a soccer ball. This design was similar to the Geodesic dome and was constructed on its structural principles. In honor of R. Buckminster Fuller, the American architect who designed this dome, this molecule was called buckminsterfullerene

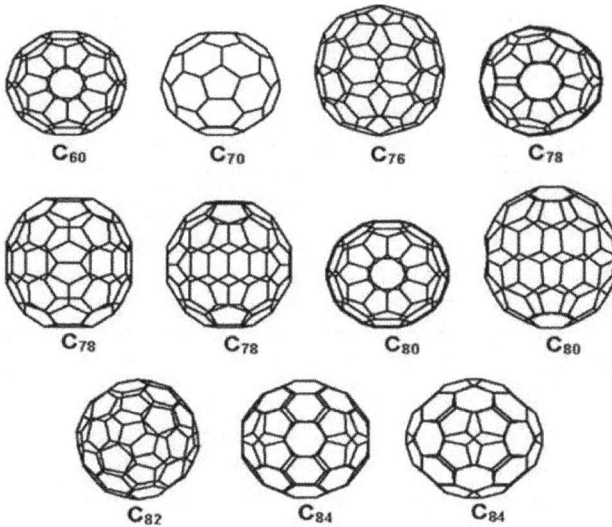

Figure 4: Fullerene structures.

or fullerene (Curl and Smalley 1991). C60 is the smallest and the most common form of fullerene. Each vertex in C60 is replaced by a pentagon. This process also converts each of the twenty former triangular faces into hexagons, with a carbon atom at the corners of each hexagon and a bond along each edge. It seems that C60 and C70 are the smallest carbon clusters for which this can be achieved. Fullerenes are similar to alkenes rather than electron-rich aromatic compounds (Kadish and Ruoff 2000). Figure 4 shows several fullerene structures.

Because of the unique physicochemical properties of these materials, fullerenes have found many applications, especially in medical fields such as photodynamic therapy, antioxidant and antiviral agents, drug delivery and gene delivery, etc. (Jensen et al. 1996). This potential and the growing application of fullerenes have brought to light several challenges about their safety and environmental effects (Colvin 2003).

Fullerenes are soluble in organic solvents and easily accept electrons. However, the solubility of fullerenes in polar solvents is known to be quite low. Some studies have shown that the solubility of fullerenes can affect their penetration into the cell and the body. For example, Scriven et al. (1994), upon synthesis of 14C-labeled fullerene, its suspension in water and its uptake by human keratinocytes, reported that it was possible to form a suspension in water that is stable for long times and can be transported to cells (Scrivens et al. 1994).

Fullerenes have strong absorption in the UV region and moderate absorption in the visible region of the electromagnetic spectrum. Available data clearly shows that pristine fullerene has no acute or sub-acute toxicity in a large diversity of living organisms, from bacteria and fungi to human leukocytes, or in drosophila, mice, rats and guinea pigs (Kolosnjaj et al. 2007). In contrast, some *in vitro* studies show that fullerenes induce cytotoxicity with increase in concentration, through a decrease in the mitochondrial membrane potential and an increase in intracellular ROS, triggering apoptosis of macrophage by activating the mitochondrial pathway (Sengul and Asmatulu 2020). Horie et al. (2013) suggest that exposure to C_{70} fullerene for 24 hours, at a concentration of 25.2 µg/mL, can lead to the induction of intracellular ROS levels in human keratinocyte and lung carcinoma cells (Horie et al. 2013). As demonstrated by an in vivo study, fullerenes can disperse in the iliac mucosa and liver at doses of 0.1, 1.0 and 10 mg/kg body weight over 92 days (Shipelin et al. 2015).

Non-functionalized fullerenes can be highly distributed in all tissues, and have been reported to have a long-lasting build up observed in bones, spleen, liver and kidneys. In concentrations above 10 mg/kg body weight, fullerenes have a toxic effect on hepatic tissues (Sengul and Asmatulu 2020).

Some modified fullerenes can lead to high toxicities (Yan et al. 2011). In a study by Xiang et al. (2012), decreased proliferation of macrophage cells, and cytotoxicity induced through the apoptosis pathway were observed in higher concentrations (100 µg/mL) of fullerene (Xiang et al. 2012).

Upon exposure to light, fullerene is an efficient singlet oxygen sensitizer. So, if pristine fullerene is absolutely nontoxic under dark conditions, this is not the case under UV-visible irradiation and in the presence of O_2, where fullerene solutions can be highly toxic through O_2 formation (Vileno et al. 2004).

3.1.2 Graphene

Graphene is a nanomaterial that was developed using a simple technique, in 2004. It is, fundamentally, a single layer of graphite, a plentiful mineral (Sharma et al. 2021) (Figure 5).

Nowadays, graphene nanoparticles (GNPs) are used as a gene transducer and biosensor for the diagnosis of diseases and tissue engineering in the healthcare sector. GNPs have also attracted much attention because of their great potential for applications in food safety and packaging, agriculture and industries (Coroş et al. 2019). However, the toxicity of graphene has become a critical concern to be addressed, despite its varied applications in multiple fields.

Several studies on the cytotoxicity of GNPs and graphene derivatives (GDs) have indicated that the most well-known mechanisms of toxicity are related to the increase in the reactive oxygen species (ROS) in cells. ROS can damage proteins, DNA and lipids, and may lead to many diseases. The release of lactate dehydrogenase (LDH) from cells, apoptosis induction

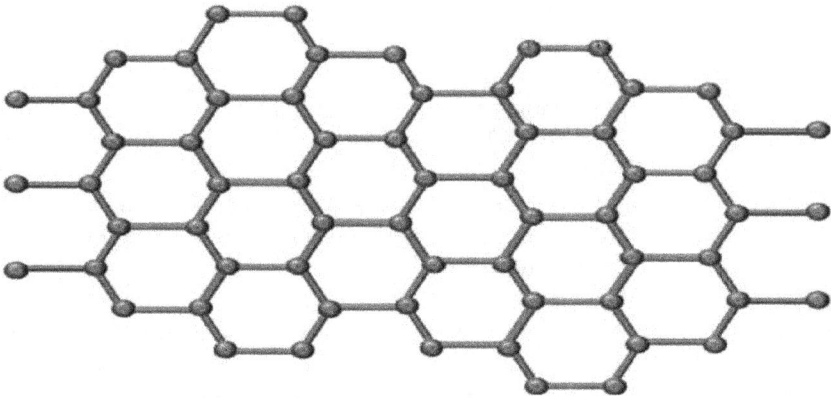

Figure 5: Structure of graphene.

and necrosis, granulomatous inflammation, and growth retardation in offsprings are also important processes that accelerate cell death by GNPs (Seabra et al. 2014). These mechanisms of toxicity of GNPs Along with sharpened edges of GDs can end in cell membrane destruction. In these mechanisms, toll-like receptors- (TLR-), transforming growth factor β- (TGF-β-) and tumor necrosis factor-alpha (TNF-α) dependent pathways are involved, and oxidative stress plays an important role therein (Ou et al. 2016, Zhang et al. 2011).

In animals, inhalation, oral administration, intravenous injection, intraperitoneal injection and subcutaneous injection are the commonly known pathways to exposure (Zhang et al. 2011). But GO-PEG and FLG do not show evident gastrointestinal tract absorption or tissue uptake via oral administration (Yang et al. 2013). Zhang et al. (2011) reported that GDs can deposit and accumulate in the mice lungs, to a high level, for more than 3 months (Zhang et al. 2011). Some studies show that graphene circulates through the body of mice in 30 minutes, and can accumulate in the liver and bladder tissues (Wen et al. 2015, Singh et al. 2011). There are evidences that show some GDs, such as GO, can agglomerate near the injection site, liver and spleen after intraperitoneal injection (Kurantowicz et al. 2015). Fu et al. (2015) believe that a low dose of GO causes main injury to the gastrointestinal system, after mice drink a GO suspension rather than a high-dose of GO; this is because a low dose of GO without agglomeration can get easily attached to the surface (Fu et al. 2015). A high concentration of aggregated-GO can block lung blood vessels and result in dyspnea. It is important to note that physicochemical properties of GDs with functional groups have a large influence on their toxicity. For example, current documents revealed no obvious pathological changes in mice exposed to low doses of GO and functionalized graphene by intravenous injection, such as aminated GO (GO-NH2), poly(acrylamide)-

functionalized GO (GO-PAM), poly (acrylic acid)-functionalized GO (GO-PAA) and GO-PEG; only GO-PEG and GO-PAA induced toxicity less than pristine GO *in vivo* (Ou et al. 2016).

Among graphene nanomaterials, carbon quantum dots, especially graphene quantum dots (GQD), are widely used because of their excellent solubility and stability in water, strong fluorescence, and the retained advantages of graphene (Wang et al. 2016). A systematic review about the *in vitro* and *in vivo* toxicity of graphene quantum dots suggests that GQD have very low cytotoxicity effects owing to their ultra-small size and high oxygen content. Also, *in vivo* studies of GQD display no material accumulation in the main organs of mice, and fast clearance of GQD through kidneys. In brief, GQD does not exhibit obvious *in vitro* and *in vivo* toxicity, even under multi-dosing situations (Chong et al. 2014).

The genotoxicity evaluation of GNPs has confirmed its carcinogenic potential in animal models. Wang et al. (2011) reported that a dose of 0.4 mg graphene oxide (GO) can cause granuloma formation in kidneys, lungs, liver and spleen, and cannot be filtered by the kidneys (Wang et al. 2011). In 2021, a systematic review study highlighted the interactions between DNA and GDs. Their genotoxicity can be classified into direct and indirect toxicity. Direct physical nucleus and DNA damage are known as direct genotoxicity mechanisms, and physical destruction, oxidative stress, epigenetic toxicity and DNA replication are indirect genotoxicity mechanisms of GDs (Wu et al. 2021).

Almost all of the studies have confirmed that cell death is significantly dependent on the dose and time (Ou et al. 2016). However, due to different size, surface modifications, etc., conflicting results are revealed with different GDs (Chatterjee et al. 2014). Most of the toxicity studies are limited to a single factor *in vitro* or *in vivo*. Hence, at present, it is impossible to arrive at any unified conclusion regarding the mechanisms of toxicity of GDs.

Recent studies have shown that when respiratory exposure to nanoparticles occurs, macrophages and neutrophils in the pulmonary system can be reactivated. In that case, it could produce an inflammatory reaction (Dudek et al. 2016). One pilot study analyzed the oxidative stress biomarkers in exhaled breath condensate (EBC) for 12 workers exposed to GDs. The biomonitoring of the exposed workers was also managed by uccal Micronucleus Cytome (BMCyt) assay, fpg-comet test (lymphocytes), and measurement of oxidized DNA bases 8-oxoGua, 8-oxoGuo and 8-oxodGuo (urine). The results of this study showed that BMCyt and fpg-comet assays are the most sensitive biomarkers, with still reparable, genotoxic and oxidative effects (Ursini et al. 2021). Although some studies have confirmed that graphene nanoparticles are present in the air of workplaces, occupational exposure limits (OELs) for them have not been recommended yet (Bellagamba et al. 2020). In 2019, a study provided the recommended occupational exposure limits (OELs) for graphene nanomaterials, based on data from a sub-chronic inhalation toxicity, using a lung dosimetry model. This study reported that the no observed adverse effect level (NOAEL) was 3.02 mg.m^{-3} and the recommended OEL by applying the uncertainty factor (UF) (an UF of 3 for species differences between rats to humans) was estimated to be 18 µg.m^{-3} (Lee et al. 2019).

In short, research on the toxicological effects of GNPs is still limited, and even occupational exposure limits (OELs) have not yet been defined for GNPs. A better understanding of the relationship between the toxicity of nanomaterials and the OEL would significantly deepen our knowledge for establishing safe workplaces.

3.1.3 Single-walled CNTs

Single-walled carbon nanotubes (SWCNTs) are the one-dimensional analogues of zero-dimensional fullerene molecules, with unique structural and electronic properties. This nanomaterial has been largely investigated as an imaging agent for the evaluation of tumor targeting (Liang and Chen 2010) (Figure 6). In fact, there is a growing trend of incorporating CNTs into industrial and consumer products ranging from advanced electronic equipment to regular household items (Maynard and Michelson 2006).

Despite such a widespread range of applications, the toxicity of CNTs is a major concern. Some studies have reported that the toxicity of CNTs is similar to that of asbestos fibers (Poland et al. 2008). In this

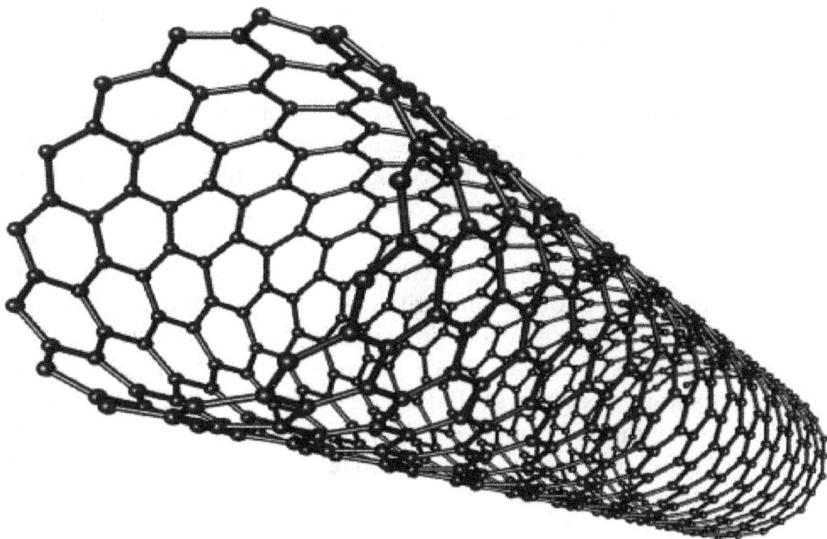

Figure 6: Structure of SWCNTs.

line, there are four main hypotheses on the mechanisms of pathogenicity of CNTs (Nagai and Toyokuni 2010). As per the first hypothesis, the generation of oxidative stress and nitrogen species (ROS/RNS) induces injury to pleural mesothelioma cells due to exposure to CNTs/asbestos fibers. Like asbestos, CNTs contain high levels of Fe, Ni, Co, Mo and other transition metal impurities. These metals or their derivatives are general components used in CNT synthesis, and all of these induce ROS/RNS formation (Vallyathan et al. 1992, Hirano et al. 2019). Also, redox-sensitive transcription factors such as AP-1 and NF-kB are activated by asbestos/ CNT exposure and can alter the expression of several genes involved in inflammation, proliferation, apoptosis and the carcinogenesis process (Karin and Marshall 1996). Several studies confirm the production of ROS by mitochondria, their localization, and the resulting damage in reaction to asbestos/CNT exposures. SWCNTs can induce oxidative damage by reducing glutathione (GSH) levels and increase malondialdehyde (MDA) levels in the liver and lungs. In brief, studies indicate that the mechanism of oxidative stress is the same for both asbestos and CNT fibers (Gupta et al. 2022).

The second theory deals with chromosome tangling due to asbestos/CNTs fibers exposure. Some studies show that, like asbestos, CNTs disrupt chromosomal distribution during mitosis, resulting in aneuploidy in the daughter cells (Oshimura et al. 1984). Sargent et al. (2012) reported mitotic spindle aberrations in cells exposed to 24, 48 and 96 $\mu g/cm^2$ SWCNTs. This finding has also been confirmed in other studies (Sargent et al. 2012). To summarize, studies obviously indicate that both asbestos and CNTs induce genotoxicity effects due to deviations in chromosomal structure, mutations and double-strand DNA breakage.

Another theory (Adsorption Theory) postulates on the surface reactivity of fibers for certain proteins and molecules. Because of the surface reactivity, carcinogenic molecules may be adsorbed on fiber surfaces and then released into the cell after fiber internalization, causing the pathogenicity (MacCorkle et al. 2006).

The last theory (Chronic Inflammation) suggests the role of continuous macrophage activation resulting in chronic inflammation (Lee et al. 2018). Several studies have showed similarities between asbestos and CNTs for inducing inflammatory reactions in human lung epithelial cells. Many inflammatory genes such as tumor necrosis factor alpha (TNF-α), the pro-inflammatory mediators NF-κB, AP-1, IL-8, IL-1β, cathepsin K, MMP12, chemokines C-C motif ligands (CCL2 and CCL3) and macrophage receptors (Toll-like receptor 2, macrophage scavenger receptor 1) are involved after the exposure of macrophages to CNTs/asbestos (Gupta et al. 2022).

Pristine SWCNT powders are lightly dispersed in air and deposited on surfaces such as exposed skin (Lam et al. 2006). Studies suggest that, after exposure to SWCNTs on *in vitro* and *in vivo* systems, there is an enhanced epidermal thickness due to the accumulation and activation of dermal fibroblasts (Murray et al. 2009). Generally, inhalation and dermal contact are the major exposure pathways of SWCNTs in the occupational setting (Ong et al. 2016).

In vivo studies such as inhalation, pharyngeal aspiration, and intratracheal instillation studies suggest that SWCNTs cause acute and chronic inflammation, granuloma formation, collagen deposition, fibrosis and genotoxic effects in the lungs (Morimoto et al. 2013). Pulmonary toxicity of induced SWCNTs is more potent than that of less dispersed ones. Inhalation exposure to SWCNTs also stimulates cardiovascular diseases in mice (Pietroiusti et al. 2009). A systematic review study has summarized the toxicity effects of SWCNTs on *in vivo* systems; oxidative stress is the most toxic effect of SWCNTs. Injected SWCNTs are distributed throughout most of the organs, including the brain, and chiefly remain in the lungs, liver and spleen, and are finally eliminated via kidneys and the bile duct. SWCNTs can be absorbed from the gastrointestinal pathway into the blood in mice and rats (Ema et al. 2016).

Occupational exposure to CNTs, specially SWCNTs, have been reported in many workplaces. However, a majority of the studies focus on the collection of area samples and industrial hygiene sampling, but very little information on toxicity upon human exposure is published.

3.1.4 Multi-walled CNTs

Multi-walled carbon nanotubes are double- and triple-walled carbon nanotubes (Figure 7). Comparing SWCNTs and MWCNTs, MWCNTs are more advantageous for reasons like low cost, ease of mass production, easy functionalization and enhanced stabilities. As a result, they have the most applications among the CNT materials, such as different kinds of electrochemical and strain-sensing applications (Nag et al. 2021).

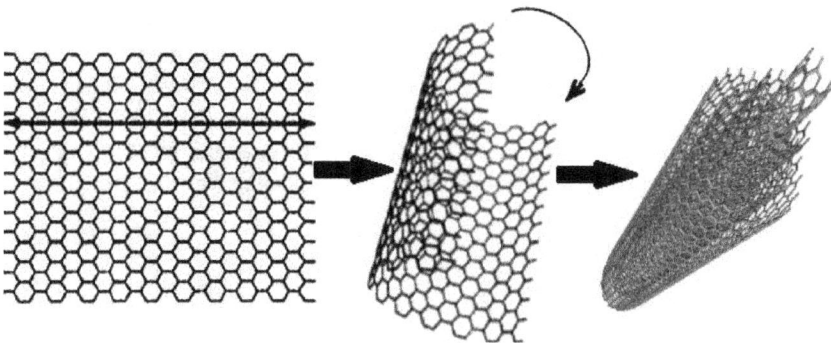

Figure 7: Structure of MWCNTs.

Toxicity studies show that inhalation is the most common pathway for MWCNTs to enter the body. In fact, MWCNTs can penetrate the pulmonary cells and cause alveolar fibrosis (Nasirzadeh et al. 2020). They have toxic effects on pulmonary epithelium. A few studies have detailed the increase in allergic reactions due to exposure to functionalized MWCNTs (F-MWCNTs) (Sharma et al. 2016). Like SWCNTs, MWCNTs are similar to asbestos and their toxicity mechanisms are based on the four hypotheses mentioned previously. However, it seems that they have higher toxicity than asbestos. One study on male Fischer rats showed that after 37–40 weeks, 6 of the 7 MWCNT-treated rats died or became moribund due to intraperitoneally disseminated mesothelioma associated with bloody ascites, while all crocidolite-treated animals survived for 52 weeks without any changes (Sakamoto et al. 2009). Based on the classification by International Agency for Research on Cancer, MWCNTs are possibly carcinogenic to humans (Group 2B) (Cancer 1997). Several studies suggest that MWCNTs can threaten the immune system and cause cardiopulmonary diseases (Poulsen et al. 2015). Zhang et al. (2017) reported that pristine-MWCNTs induce a remarkable degradation in lymphatic cells (Zhang et al. 2017). It should be mentioned that modified MWCNTs can cause different toxic effects and we are talking about the major and general effects of MWCNTs here.

Some studies have compared the toxicity of MWCNTs with other CNTs. For example, Mohammadian et al. (2019) reported that MWCNTs are more toxic than SWCNTs and the toxicology indicators for MWCNTs are less than those of SWCNTs (Mohammadian et al. 2019). In contrast, some studies suggest that SWCNTs are toxic to organs and induce death at high dosages (Kavosi et al. 2018). It is clear that the different physicochemical properties of CNTs are the reason for these disagreements.

Because of the conflicting results regarding a potential hazard and the toxicity of CNTs, different occupational exposure limits (OEL) have been proposed. The National Institute for Occupational Safety and Health (NIOSH) has established that the recommended exposure limit (REL) for CNTs should be 1 $\mu g.m^{-3}$ during a 40-hour workweek, 50 weeks per year for 45 years (Fonseca et al. 2015). In some studies, OELs have been recommended in the range of 1–50 $\mu g\ m^{-3}$ (8-h TWA) (Myojo and Ono-Ogasawara 2018). Lee et al. (2018) estimated that the OEL for MWCNTs was 6 $\mu g.m^{-3}$ (Lee et al. 2019). Therefore, achieving the OEL for CNTs needs complete information on human exposure and toxic effects.

3.2 Metal or Metal Oxide Nanomaterials

3.2.1 Gold

There are conflicting outcomes regarding the toxicity of gold nanoparticles. Some studies report that these particles are biocompatible and nontoxic (Mukherjee et al. 2016), whereas others have reported that they induce oxidative stress and have high toxic effects (Lopez-Chaves et al. 2018).

The results of studies show that gold nanoparticles with positively charged coatings, such as branched polyethylenimine or amine terminated (silicon and polystyrene nanoparticles), have the main toxic effects in Caenorhabditis elegans (Sengul and Asmatulu 2020). Interestingly, in a study by Steckiewicz et al. (2019), gold nanoparticle stars ($\approx$ 215 nm) with the highest anticancer potential had the most cytotoxic effects, while nanoparticle spheres ($\approx$ 6.3 nm) with the lowest anticancer potential had the least toxic effects (Steckiewicz et al. 2019).

Thota and Crans (2018) showed that if the gold nanoparticles are taken up by a healthy cell, it can be removed by endosomes/lysosomes, whereas the result will be cell death if it is taken up by a cancer cell (Thota and Crans 2018). A systematic review showed that short-term exposure to gold nanoparticles may not be toxic to cells, whereas long-term exposure may disturb cellular metabolism as well as energy homeostasis (Sengul and Asmatulu 2020).

3.2.2 Silver

Because of the antibacterial, antifungal and antiviral properties of silver nanoparticles, the use of this material for biomedical purposes such as antimicrobial agents, drug delivery, molecular imaging, biomedical sensing and even cancer photodynamic therapy has increased (Cameron et al. 2018). However, the toxicological effects of silver nanoparticles have been reported in many studies (Sengul and Asmatulu 2020).

Silver nanoparticles can enter the body via different pathways and accumulate in various tissues and organs, then pass the blood–brain barrier and finally reach the brain cells (Ciucă et al. 2017). These nanoparticles can induce ROS and leakage of the enzyme lactate dehydrogenase (LDH) and lead to the most grave of toxicity effects such as necrosis (Dasgupta and Ramalingam 2016). The results of some studies show that the smallest sized silver nanoparticles display a greater ability to induce hemolysis and membrane damage than bulk particles (Guo et al. 2019).

3.2.3 Cobalt

Cobalt-based nanoparticles are used in pigments, catalysts, sensors, magnetic contrast agents and energy storage devices (Cappellini et al. 2018).

The potential toxic effects of cobalt-based nanoparticles have been reported previously in *in vivo* and *in vitro* studies. The results of these studies show that exposure to cobalt-based nanoparticles induces oxidative stress, DNA damage, morphological transformation and inflammatory reactions in human cells (Wan et al. 2017). Pure cobalt oxide nanoparticles can induce cell death by producing ROS, which induces tumor necrosis and leads to toxic effects on human immune cells (Wan et al. 2017).

Some studies have reported that cobalt and cobalt (II) oxide nanoparticles can create genotoxic effects in mice. Genotoxic effects appear with oxidative stress, lung inflammation and injury, and cell proliferation, which further results in DNA damage and mutation (Cappellini et al. 2018).

3.2.4 Aluminum Oxide

Aluminum oxide is used in paints, polymers, coatings, textiles, fuel cells, solar energy, airbag propellants and energy materials (Poborilova et al. 2013).

Some studies show that exposure to aluminum oxide nanoparticles may lead to decreased cell viability, increased oxidative stress, mitochondrial dysfunction and change in the protein expression of the blood–brain barrier (Alshatwi et al. 2012; Kim et al. 2010; Pakrashi et al. 2011).

Aluminum oxide nanoparticles can also induce cytotoxic effects and membrane damage in animal cells that may be mediated through morphological abnormalities, lipid peroxidation and oxidative stress (Srikanth et al. 2015).

3.2.5 Titanium Dioxide

Titanium dioxide is one of the top five nanoparticles registered in consumer products (Gea et al. 2019). They have been extensively used in sunscreens, cosmetics, toothpaste, pharmaceuticals, paints, plastics, self-cleaning devices and food additives, and in industrial and medical sectors (Ursini et al. 2014).

There are conflicting studies on the toxic effects of titanium oxide nanoparticles investigated both in vitro and in vivo models (Sengul and Asmatulu 2020). Some studies show that these nanoparticles are not toxic (Hobbs et al. 2016), and the cytotoxicity is low overall and influenced by the nanoparticles' shape as well as light exposure. Besides, no significant lactate dehydrogenase release is detected (Petersen et al. 2014). Other studies suggest that titanium dioxide nanoparticles can induce cytotoxic, genotoxic and oxidative effects through oxidant generation, inflammation and apoptosis (Proquin et al. 2017). Different studies show that the presence of light enhances the genotoxic effects of some nanoparticles, mainly increasing the oxidative stress (Mohammadian and Nasirzadeh, 2021). In this line, the phototoxicity of titanium dioxide nanoparticles of four different sizes (< 25 nm, 31 nm, < 100 nm and 325 nm) and two crystal forms (anatase and rutile) towards human skin keratinocytes under ultraviolet radiation has been evaluated by researchers. The results showed that ROS was produced by ultraviolet radiation of all the titanium dioxide nanoparticles. Interestingly, smaller titanium dioxide nanoparticles resulted in higher phototoxicity (Xin et al. 2019).

Titanium dioxide nanoparticles are classified as a possible carcinogen for humans, based on the International Agency for Research on Cancer (Cancer 2010). Also, the National Institute of Occupational Safety and Health recommends exposure limits of 0.3 mg/m^3 for ultrafine titanium dioxide (Baker 2012).

3.2.6 Platinum

Platinum nanoparticles are mainly used as therapeutic agents in medicine for the treatment of cancer cells. They have potent anticancer activities against ovarian teratocarcinoma cell line via induction of apoptosis and cell cycle arrest (Bendale et al. 2017). Many studies have proved that platinum nanoparticles are not toxic or have low toxicity effects. For example, Samadi et al. (2018) showed that platinum nanoparticles did not cause changes in cellular oxidative stress or apoptotic or necrotic cell death after 24 hours of exposure at different concentrations (0–50 µg/ ml), on living cells (Samadi et al. 2018). One study shows that platinum nanoparticles cause toxic effects on primary keratinocytes, decreasing cell metabolism; however, these changes have no effects on cell viability (Konieczny et al. 2013).

On the contrary, some investigations suggest that platinum nanoparticles can induce cytotoxic effects on human lung adenocarcinoma, ovarian teratocarcinoma, pancreatic cancer cells and normal peripheral blood mononucleocyte cells (Bendale et al. 2017).

Overall, toxicity of platinum nanoparticles depends on their size. In fact, smaller nanoparticles exhibit more deleterious effects on DNA stability and higher caspases activation, compared to bulk particles (Konieczny et al. 2013). Also, findings on dose-dependent cytotoxicity show that a high dosage of platinum nanoparticles induces cytotoxicity in human liver cells, but a low-level dosage can elicit multiple stress reactions, secretion of proinflammatory cytokines and modulation of insulin-like growth factor-1-dependent signal transduction (Bendale et al. 2017).

3.2.7 Zinc Oxide

Zinc oxide nanoparticles are broadly used in catalysis, paints, wave filters, UV detectors, transparent conductive films, varistors, gas sensors, solar cells, sunscreens, cosmetic products and UV blockers (Subramaniam et al. 2019). Due to their wide application in cosmetics and daily care products and as a result of their excessive consumption, exposure to them have increased (Senapati and Kumar 2018).

However, many studies have reported that zinc oxide nanoparticles are not toxic, whereas some investigations show the toxic effects of zinc oxide, such as decrease in cell viability, remarkable morphological changes, apoptosis induction via ROS production and IL-8 release after 24 hours of human colon carcinoma cells' exposure to zinc oxide (5 $\mu g/cm^2$) (De Berardis et al. 2010). Cytotoxic effects are also observed on hepatocyte and human embryonic kidney cells (Guan et al. 2012).

A review study shows the toxicological profiles of zinc oxide nanoparticles in human lung fibroblasts in *in vitro* and *in vivo* models. For *in vitro* toxicity, the results suggested a significant release of extracellular lactate dehydrogenase and decreased cell viability. Also, oxidative stress-induced DNA damage was defined by the release of a DNA oxidation product and 8-hydroxydeoxyguanosine (Ng et al. 2017). For *in vivo* study, oxidative stress associated with cytotoxicity and genotoxicity, elevated level of malondialdehyde and decreased the level of superoxide dismutase were observed in human lung fibroblasts, and kidney cells (Xiao et al. 2016, Sengul and Asmatulu 2020).

3.2.8 Iron Oxide

Iron oxide nanoparticles have been broadly used for biomedical applications including drug delivery, magnetic resonance imaging and thermal ablation therapy (Feng et al. 2018).

Some investigations suggest that inhalation of iron oxide nanoparticles can cause oxidative stress in the liver, spleen, lungs and brain, which results in inflammation, low cell viability, cell lysis and disturbance of the blood coagulation system (Tan et al. 2018). Another study shows an increase in oxidative damage, apoptosis and reduced cell viability after 12-hour and 24-hour exposure to 25, 50, 75 and 100 $\mu g/ml$ of iron oxide nanoparticles (Sadeghi et al. 2015).

Magdolenova et al. (2015) reported that uncoated iron oxide nanoparticles are not found to be cytotoxic or genotoxic, while oleate-coated iron oxide nanoparticles are found to be cytotoxic in a

Table 1: Toxic mechanisms of critical metals or metal oxide nanomaterials.

Type of metal	Toxicity effects
Gold	oxidative stress.
Silver	oxidative stress, necrosis.
Cobalt-based nanomaterials	oxidative stress, necrosis, inflammatory reactions, genotoxic effects.
Aluminum oxide	oxidative stress, cell death, mitochondrial dysfunction.
Titanium dioxide	oxidative stress, lactate dehydrogenase release, phototoxicity.
Platinum	oxidative stress, decreasing cell metabolism, apoptosis, changing DNA stability, secretion of pro-inflammatory cytokines.
Zinc oxide	oxidative stress, cell death, apoptosis, morphological changes, DNA damage, secretion of pro-inflammatory cytokines, release of extracellular ldh, elevated level of malondialdehyde, decreased the level of superoxide dismutase.
Iron oxide	oxidative stress, decreasing cell viability, apoptosis, DNA damage.

dose-dependent manner and also induce DNA damage, indicating their genotoxic potential (Magdolenova et al. 2015). In this line, studies have demonstrated that both size and coating have a remarkable effect on cytotoxicity. Polyethylenimine-coated iron oxide nanoparticles have been found to exhibit higher cytotoxicity than polyethylene glycol-coated iron oxide nanoparticles. Also, smaller-sized polyethylene glycol-coated iron oxide nanoparticles (10 nm) display fairly higher cellular uptake than larger ones (30 nm) (Feng et al. 2018).

Toxic mechanisms of the aforementioned metals or metal oxide nanomaterials are summarized in Table 1.

3.3 Other Nanomaterials

Silica Nanoparticles

Silicon dioxide nanoparticles are amorphous materials, generally spherical in shape (Figure 8). They can be made to have a broad range of sizes and their surface chemistry can be easily modified to target a variety of applications. Due to their novel properties, silica nanoparticles are widely used in agriculture, food, consumer products (including cosmetics) and bio-medical applications (such as gene carrier, drug delivery and molecular imaging) (Brinch et al. 2016, Bitar et al. 2012). In 2013, silica nanoparticles became one of the three most produced nanomaterials worldwide (Ryu et al. 2014).

Figure 8: Structure of silica nanoparticles.

In recent years, due to the global commercialization of silica nanoparticles, the risk of human exposure in workplaces has increased (Kim et al. 2014). Also, the use of silica nanoparticles in medical applications can be intentionally introduced into the human body for disease diagnosis and treatments, and aggravate the potential adverse health effects (Croissant et al. 2017).

Like other nanomaterials, physicochemical characteristics such as size, surface area and surface features play a key role in the toxicity of silica nanoparticles as well (Napierska et al. 2010). The toxicological potential of silica has been related to its crystallinity. Current studies have shown that amorphous silica nanoparticles can be as reactive as crystalline particles (Ryu et al. 2014), but *in vivo* studies show that amorphous silica nanoparticles are cleared more easily from the lungs, and may thus have a lower pathogenic potential (Arts et al. 2007).

Studies have stated that exposure to silica nanoparticles can induce toxic effects *in vitro* (immortalized mammalian cell lines) and *in vivo* (rats and mice). In *in vitro* systems, induction of oxidative stress is considered as the major mechanism involving toxicity. ROS production, lipid peroxidation, lactate dehydrogenase (LDH) release (loss of membrane integrity), etc. have been observed as the mechanisms of toxicity (Murugadoss et al. 2017). A systematic review of the genotoxicity of silica nanoparticles showed that interaction with DNA, oxidative DNA damage, depletion of anti-oxidants, cell cycle arrest and abnormal expression of genes were identified as the potential toxicity mechanisms (Yazdimamaghani et al. 2019). According to immunotoxicity studies, not only do silica nanoparticles induce stronger pro-inflammatory responses compared to bulk materials, but the size effect as well is an important factor in immunotoxicity (McCullen et al. 2007).

In vivo toxicity studies show that silica nanoparticles penetrate mainly via inhalation, ingestion and intravenous routes, are majorly distributed in the liver, lungs, spleen, kidneys, and in the brain of exposed rats, especially with short-term exposure, and induce adverse effects in these organs (Patel and Patel 2015). Some studies show the accumulation of silica nanoparticles in organs such as the liver, but such accumulation was not linked to any major effects. Interestingly, silica nanoparticles show a significant increase in absorption by the gastrointestinal tract, compared to pure nanomaterials (Murugadoss et al. 2017).

In contrast to acute studies, the long-term effects of accumulated silica nanoparticles have not been studied. No local or systemic toxicity has been observed in chronic oral and dermal exposure studies (Croissant et al. 2020).

Based on the occupational exposure levels (OELs), Murugadoss et al. (2017) estimated that doses as high as 384 $\mu g/cm^2$ for human inhalation exposure to amorphous silica are not toxic (Murugadoss et al. 2017).

4. Controlling Exposure to Nanoparticle

The health effects associated with nanomaterials are not yet clearly understood; so, it is important for producers and users of nanomaterials to reduce employees' exposure appropriately. Exposure control is the use of strategies or a set of tools for eliminating or decreasing the exposure of workers to nanomaterials. In 2013, the National Institute for Occupational Safety and Health (NIOSH) published a guideline of control approaches for nanomaterial production and use processes. This guideline provides solutions towards exposure control for protecting workers during the handling of nanomaterials. Exposure control consists of a standardized hierarchy, including elimination, substitution, isolation, engineering controls, administrative controls and personal protective equipment (PPE) (Figure 9).

4.1 Elimination

Elimination is generally not practicable for workers handling nanomaterials. However, it may be possible to change certain aspects of the physical form of the nanomaterial or the process in a way that decreases nanomaterial distribution.

4.2 Substitution

Like controlling exposure by elimination, substitution is not often feasible for workers handling nanomaterials.

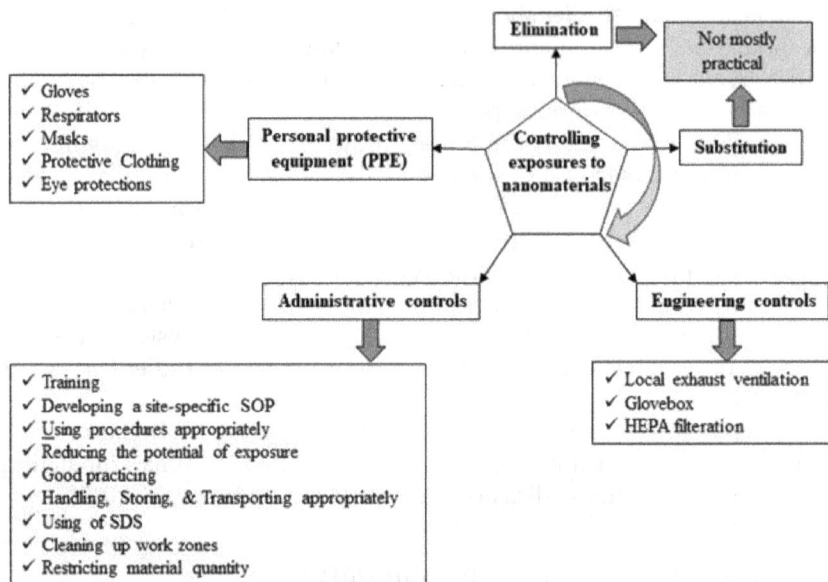

Figure 9: Hierarchy of controls for activities involving nanomaterials.

4.3 *Engineering Controls*

Engineering controls used in nanotechnology applications are likely to be similar to those currently used in controlling aerosols (gases, dusts, chemical vapors, etc.) found in other laboratory applications and/or processes. Such controls may include:

- Local exhaust ventilation (LEV), such as the standard laboratory chemical hood
    - Class II, Type B2 Biosafety Cabinet (BSC) [no recirculation]
    - Class II, Type A2 or Class II, Type B1 BSC
    - Glovebox
- Filtration by a high-efficiency particulate air (HEPA) filter

    The use of laminar-flow clean benches is not suggested for the control of nanoparticles.

4.4 *Administrative Controls*

Administrative controls include training, procedure, policy, shift designs, etc. that reduce the threats of nanomaterials for workers. Good work practices involving nanoparticles should include awareness of the following practices:

- Training: Every person working with nanomaterials must have an up-to-date training on nanomaterial safety. Workplace-specific training for the specific classes of nanomaterials, along with the specific way, must be provided. Training can be informal, such as with a demonstration, or verbal or written instructions, and can be delegated to a representative of the manager.
- Developing a site-specific standard operating procedure (SOP) for work involving nanoparticles.
- Reducing the potential for inhalation exposure and skin contact.
- Practicing good personal hygiene (e.g., hand washing, etc.).
- Appropriately following procedures when using laboratory equipment.
- Following the manufacturer or vendor instructions for the use or handling of nanoparticles.
- Handling, storing and transporting nanoparticles (in liquid or powder state) in closed, sealed and labeled containers.
- Using manufacturer supplied Material Safety Data Sheet (MSDS) for specific safety and health information.
- Applying wet-wiping methods to clean up work zones, using a solvent expected to solubilize the nanoparticle in use.
- Restricting material quantity to what is needed.

4.5 Personal Protective Equipment (PPE)

The priority of control measures is always given to the first levels. Because of many uncertainties regarding the potential hazards of nanomaterials, most of the control approaches are focused on the last three levels of the hierarchy of control (engineering controls, administrative controls and PPE) (Freeland et al. 2016). Although personal protective equipment (PPE) is the last priority of controls, investigations suggest that the use of engineering and administrative controls do not provide the essential control, and suitable PPEs for working with nanomaterials should be used to complete the protection (Freeland et al. 2016, Landsiedel et al. 2017). There are limited referenced guidelines for suitable PPEs protecting against nanoparticles. However, some guides for suitable PPEs have been recommended when working with nanoparticles. In Table 2, suitable PPEs for working with nanomaterials have been summarized.

Table 2: Suitable PPEs for working with nanomaterials.

PPEs	Type of PPEs	Descriptions
gloves	latex nitrile neoprene vinyl	Wearing two layers of gloves is the best practice for preventing nanoparticle penetration into skin.
respirators	half-face respirators full-face respirators	• Size of the nanoparticle should be evaluated for determining the appropriate respirator. • An OEL may exist for a particular type of nanoparticle. Determining the need for a respirator should also include a review of these aOELs.
masks	N95 N100 FFP3 FFP2	Dust masks and surgical masks should not be used.protective clothing
protective clothing	closed-toed shoes long pants without cuff long shirts with sleeves lab coats (non-cotton)	• These clothing are appropriate for wet-chemistry labs. • After using the clothing and before taking them out of the laboratory, they should be placed in closed bags.
eye protection	safety glasses safety goggles face shields	Face shield alone is not sufficient protection against unbound dry materials.

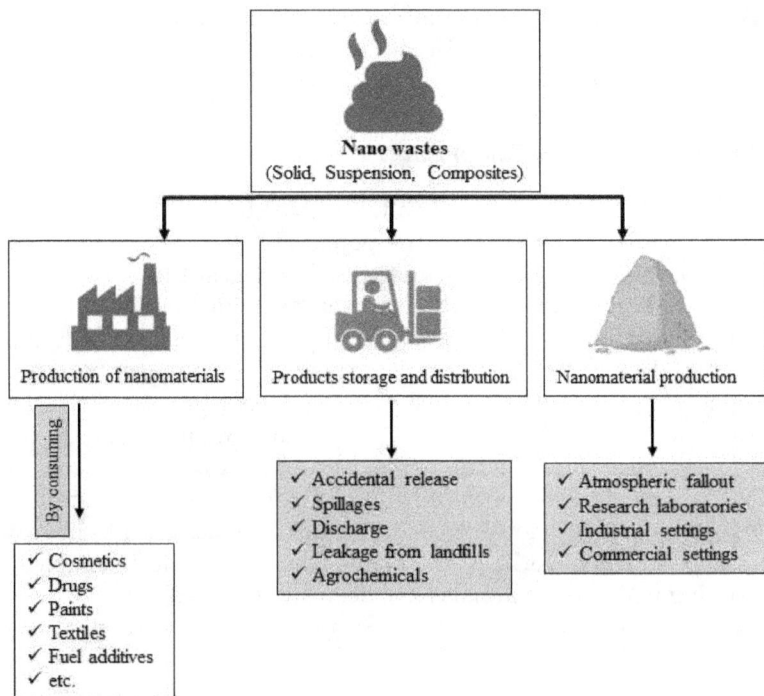

Figure 10: Nanowastes.

5. Nanomaterial Waste Management

Nanomaterial wastes may stream into the waste system from various sources such as industries, households, hospitals, etc. Most nanoparticles in the industry contain heavy metals that are known for their critical environmental concerns. During the process of nanomaterial waste management, the effects of new materials on the environment need to be evaluated because of their possible effects on biological systems. One of the key matters is the proliferation of nano-wastes, which still remains almost unknown. Solving these problems requires new eco-friendly awareness and guidance with regard to the legalization of risk-inducing nanomaterials. However, we should consider that all nanomaterials must be disposed as chemical wastes. Powdered materials or dispersions should be double bagged and disposed of in rigid, sealed and labelled containers, a green pail labelled 'nanomaterial waste' for instance. Also, nanomaterial waste streams should not be mixed (Figure 10).

References

Alhamoud, M., Yi, C.X., Alghamdi, M.A. et al. 2022. Biological toxicity and environmental hazards associated with polymeric micelles. Polymeric Micelles for Drug Delivery, 593–628.

Alshatwi, A.A., Vaiyapuri Subbarayan, P., Ramesh, E. et al. 2012. Al2O3 nanoparticles induce mitochondria-mediated cell death and upregulate the expression of signaling genes in human mesenchymal stem cells. Journal of Biochemical and Molecular Toxicology 26(11): 469–476.

Aqel, A., Abou El-Nour, K.M., Ammar, R.A. et al. 2012. Carbon nanotubes, science and technology part (I) structure, synthesis and characterisation. Arabian Journal of Chemistry 5(1): 1–23.

Arts, J.H., Muijser, H., Duistermaat, E. et al. 2007. Five-day inhalation toxicity study of three types of synthetic amorphous silicas in Wistar rats and post-exposure evaluations for up to 3 months. Food and Chemical Toxicology 45(10): 1856–1867.

Baker, J. 2012. Current Issues in Exposure Assessment for Nanoparticles. ASSE Professional Development Conference and Exposition. OnePetro.

Banoun, H. 2022. Current state of knowledge on the excretion of mRNA and spike produced by anti-COVID-19 mRNA vaccines; possibility of contamination of the entourage of those vaccinated by these products. Infect. Dis. Res. 3(4): 22.

Basinas, I., Jiménez, A.S., Galea, K.S. et al. 2018. A systematic review of the routes and forms of exposure to engineered nanomaterials. Annals of Work Exposures and Health 62(6): 639–662.

Bellagamba, I., Boccuni, F., Ferrante, R. et al. 2020. Workers' exposure assessment during the production of graphene nanoplatelets in R&D laboratory. Nanomaterials 10(8): 1520.

Bendale, Y., Bendale, V. and Paul, S. 2017. Evaluation of cytotoxic activity of platinum nanoparticles against normal and cancer cells and its anticancer potential through induction of apoptosis. Integrative Medicine Research 6(2): 141–148.

Bergamaschi, E. 2009. Occupational exposure to nanomaterials: present knowledge and future development. Nanotoxicology 3(3): 194–201.

Bitar, A., Ahmad, N.M., Fessi, H. et al. 2012. Silica-based nanoparticles for biomedical applications. Drug Discovery Today 17(19-20): 1147–1154.

Borm, P., Klaessig, F.C., Landry, T.D. et al. 2006. Research strategies for safety evaluation of nanomaterials, Part V: Role of dissolution in biological fate and effects of nanoscale particles. Toxicological Sciences 90(1): 23–32.

Brinch, A., Hansen, S.F., Hartmann, N.B. et al. 2016. EU regulation of nanobiocides: Challenges in implementing the biocidal product regulation (BPR). Nanomaterials 6(2): 33.

Bruinink, A., Wang, J. and Wick, P. 2015. Effect of particle agglomeration in nanotoxicology. Archives of Toxicology 89(5): 659–675.

Cameron, S.J., Hosseinian, F. and Willmore, W.G. 2018. A current overview of the biological and cellular effects of nanosilver. International Journal of Molecular Sciences 19(7): 2030.

Cancer IAfRo 1997. Evaluation of carcinogenic risks to humans. Silica, some silicates, coal dust, and para-aramid fibrils 68.

Cancer IAfRo 2010. Carbon black, titanium dioxide, and talc. IARC Press, International Agency for Research on Cancer.

Canu, I.G., Bateson, T.F., Bouvard, V. et al. 2016. Human exposure to carbon-based fibrous nanomaterials: A review. International Journal of Hygiene and Environmental Health 219(2): 166–175.

Cappellini, F., Hedberg, Y., McCarrick, S. et al. 2018. Mechanistic insight into reactivity and (geno) toxicity of well-characterized nanoparticles of cobalt metal and oxides. Nanotoxicology 12(6): 602–620.

Chatterjee, N., Eom, H.-J. and Choi, J. 2014. A systems toxicology approach to the surface functionality control of graphene–cell interactions. Biomaterials 35(4): 1109–1127.

Chen, Z., Yadghar, A.M., Zhao, L. and Mi, Z. 2011. A review of environmental effects and management of nanomaterials. Toxicological & Environmental Chemistry 93(6): 1227–1250.

Chong, Y., Ma, Y., Shen, H. et al. 2014. The *in vitro* and *in vivo* toxicity of graphene quantum dots. Biomaterials 35(19): 5041–5048.

Christian, P., Von der Kammer, F., Baalousha, M. et al. 2008. Nanoparticles: Structure, properties, preparation and behaviour in environmental media. Ecotoxicology 17(5): 326–343.

Ciucă, A.G., Grecu, C.I., Rotărescu, P. et al. 2017. Nanostructures for drug delivery: Pharmacokinetic and toxicological aspects. Nanostructures for Drug Delivery. Elsevier, pp. 941–957.

Colvin, V.L. 2003. The potential environmental impact of engineered nanomaterials. Nature Biotechnology 21(10): 1166–1170.

Coroş, M., Pogăcean, F., Măgeruşan, L. et al. 2019. A brief overview on synthesis and applications of graphene and graphene-based nanomaterials. Frontiers of Materials Science 13(1): 23–32.

Council, N.R. 2011. Prudent practices in the laboratory: Handling and management of chemical hazards, updated version.

Croissant, J.G., Butler, K.S., Zink, J.I. et al. 2020. Synthetic amorphous silica nanoparticles: Toxicity, biomedical and environmental implications. Nature Reviews Materials 5(12): 886–909.

Croissant, J.G., Fatieiev, Y. and Khashab, N.M. 2017. Functional nanoparticles: degradability and clearance of silicon, organosilica, silsesquioxane, silica mixed oxide, and mesoporous silica nanoparticles (Adv. Mater. 9/2017). Advanced Materials 29(9).

Curl, R.F. and Smalley, R.E. 1991. Fullerenes. Scientific American 265(4): 54–63.

Dasgupta, N. and Ramalingam, C. 2016. Silver nanoparticle antimicrobial activity explained by membrane rupture and reactive oxygen generation. Environmental Chemistry Letters 14(4): 477–485.

De Berardis, B., Civitelli, G., Condello, M. et al. 2010. Exposure to ZnO nanoparticles induces oxidative stress and cytotoxicity in human colon carcinoma cells. Toxicology and Applied Pharmacology 246(3): 116–127.

De Jong, W.H. and Borm, P.J. 2008. Drug delivery and nanoparticles: Applications and hazards. International Journal of Nanomedicine 3(2): 133.

Diao, J., Hua, D., Lin, J. et al. 2005. Nanoparticle delivery by controlled bacteria. Journal of Nanoscience and Nanotechnology 5(10): 1749–1751.

Dong, J. and Ma, Q. 2019. Integration of inflammation, fibrosis, and cancer induced by carbon nanotubes. Nanotoxicology 13(9): 1244–1274.

Dreizin, E.L. 2009. Metal-based reactive nanomaterials. Progress in Energy and Combustion Science 35(2): 141–167.

Dudek, I., Skoda, M., Jarosz, A. et al. 2016. The molecular influence of graphene and graphene oxide on the immune system under *in vitro* and *in vivo* conditions. Archivum Immunologiae et Therapiae Experimentalis 64(3): 195–215.

Eckhoff, R. 2003. Dust Explosions in the Process Industries: Identification, Assessment and Control of Dust Hazards. Elsevier.

Eckhoff, R.K. 2013. Influence of dispersibility and coagulation on the dust explosion risk presented by powders consisting of nm-particles. Powder Technology 239: 223–230.

Ema, M., Gamo, M. and Honda, K. 2016. A review of toxicity studies of single-walled carbon nanotubes in laboratory animals. Regulatory Toxicology and Pharmacology 74: 42–63.

Feng, Q., Liu, Y., Huang, J. et al. 2018. Uptake, distribution, clearance, and toxicity of iron oxide nanoparticles with different sizes and coatings. Scientific Reports 8(1): 1–13.

Florence, A.T. 1997. The oral absorption of micro-and nanoparticulates: Neither exceptional nor unusual. Pharmaceutical Research 14(3): 259–266.

Fonseca, A.S., Viitanen, A.-K., Koivisto, A.J. et al. 2015. Characterization of exposure to carbon nanotubes in an industrial setting. Annals of Occupational Hygiene 59(5): 586–599.

Freeland, J., Hulme, J., Kinnison, D., Mitchell, A., Veitch, P., Aitken, R. et al. 2016. Working safely with nanomaterials in research & development. Reportno. Report Number I, Date. Place Published I: Institution I

Fu, C., Liu, T., Li, L. et al. 2015. Effects of graphene oxide on the development of offspring mice in lactation period. Biomaterials 40: 23–31.

Gea, M., Bonetta, S., Iannarelli, L. et al. 2019. Shape-engineered titanium dioxide nanoparticles (TiO2-NPs): Cytotoxicity and genotoxicity in bronchial epithelial cells. Food and Chemical Toxicology 127: 89–100.

Ghafari, J., Moghadasi, N. and Shekaftik, S.O. 2020. Oxidative stress induced by occupational exposure to nanomaterials: A systematic review. Industrial Health 58(6): 492–502.

Ghumman, M., Dhamecha, D., Gonsalves, A. et al. 2021. Emerging drug delivery strategies for idiopathic pulmonary fibrosis treatment. European Journal of Pharmaceutics and Biopharmaceutics 164: 1–12.

Gimbert, L.J., Hamon, R.E., Casey, P.S. et al. 2007. Partitioning and stability of engineered ZnO nanoparticles in soil suspensions using flow field-flow fractionation. Environmental Chemistry 4(1): 8–10.

Gimeno-Benito, I., Giusti, A., Dekkers, S. et al. 2021. A review to support the derivation of a worst-case dermal penetration value for nanoparticles. Regulatory Toxicology and Pharmacology 119: 104836.

Gomez-Villalba, L.S., Salcines, C. and Fort, R. 2023. Application of inorganic nanomaterials in cultural heritage conservation, risk of toxicity, and preventive measures. Nanomaterials 13(9): 1454.

Guan, R., Kang, T., Lu, F. et al. 2012. Cytotoxicity, oxidative stress, and genotoxicity in human hepatocyte and embryonic kidney cells exposed to ZnO nanoparticles. Nanoscale Research Letters 7(1): 1–7.

Guldi, D.M., Huie, R.E., Neta, P. et al. 1994. Excitation of C60, solubilized in water by triton X-100 and γ-cyclodextrin, and subsequent charge separation via reductive quenching. Chemical Physics Letters 223(5-6): 511–516.

Guo, Z., Zeng, G., Cui, K. et al. 2019. Toxicity of environmental nanosilver: Mechanism and assessment. Environmental Chemistry Letters 17(1): 319–333.

Gupta, S.S., Singh, K.P., Gupta, S. et al. 2022. Do carbon nanotubes and asbestos fibers exhibit common toxicity mechanisms? Nanomaterials 12(10): 1708.

Hirano, A., Wada, M., Tanaka, T. et al. 2019. Oxidative stress of carbon nanotubes on proteins is mediated by metals originating from the catalyst remains. ACS Nano 13(2): 1805–1816.

Hobbs, C.A., Davis, J., Shepard, K. et al. 2016. Differential genotoxicity of acrylamide in the micronucleus and Pig-a gene mutation assays in F344 rats and B6C3F1 mice. Mutagenesis 31(6): 617–626.

Hodson, L., Methner, M. and Zumwalde, R.D. 2009. Approaches to safe nanotechnology; managing the health and safety concerns associated with engineered nanomaterials.

Horie, M., Nishio, K., Kato, H. et al. 2013. *In vitro* evaluation of cellular influences induced by stable fullerene C70 medium dispersion: Induction of cellular oxidative stress. Chemosphere 93(6): 1182–1188.

Iavicoli, I., Fontana, L., Pingue, P. et al. 2018. Assessment of occupational exposure to engineered nanomaterials in research laboratories using personal monitors. Science of the Total Environment 627: 689–702.

Iavicoli, I., Leso, V., Manno, M. et al. 2014. Biomarkers of nanomaterial exposure and effect: current status. Journal of Nanoparticle Research 16(3): 1–33.

Jensen, A.W., Wilson, S.R. and Schuster, D.I. 1996. Biological applications of fullerenes. Bioorganic & Medicinal Chemistry 4(6): 767–779.

Kadish, K.M. and Ruoff, R.S. 2000. Fullerenes: Chemistry, Physics, and Technology. John Wiley and Sons.

Karin, M. and Marshall, C.J. 1996. The regulation of AP-1 activity by mitogen-activated protein kinases. Philosophical Transactions of the Royal Society of London. Series B: Biological Sciences 351(1336): 127–134.

Kavosi, A., Hosseini Ghale Noei, S., Madani, S. et al. 2018. The toxicity and therapeutic effects of single-and multi-wall carbon nanotubes on mice breast cancer. Scientific Reports 8(1): 1–12.

Khan, F.H. 2013. Chemical hazards of nanoparticles to human and environment (a review). Oriental Journal of Chemistry 29(4): 1399.

Kim, B., Kim, H. and Yu, I.J. 2014. Assessment of nanoparticle exposure in nanosilica handling process: Including characteristics of nanoparticles leaking from a vacuum cleaner. Industrial Health 52(2): 152–162.

Kim, I.-S., Baek, M. and Choi, S.-J. 2010. Comparative cytotoxicity of Al2O3, CeO2, TiO2 and ZnO nanoparticles to human lung cells. Journal of Nanoscience and Nanotechnology 10(5): 3453–3458.

Kis, N., Kovács, A., Budai-Szűcs, M. et al. 2022. The effect of non-invasive dermal electroporation on skin barrier function and skin permeation in combination with different dermal formulations. Journal of Drug Delivery Science and Technology 69: 103161.

Klaine, S.J., Alvarez, P.J., Batley, GE. et al. 2008. Nanomaterials in the environment: Behavior, fate, bioavailability, and effects. Environmental Toxicology and Chemistry: An International Journal 27(9): 1825–1851.

Kolosnjaj, J., Szwarc, H. and Moussa, F. 2007. Toxicity studies of fullerenes and derivatives. Bio-Applications of Nanoparticles, 168–180.

Konieczny, P., Goralczyk, A.G., Szmyd, R. et al. 2013. Effects triggered by platinum nanoparticles on primary keratinocytes. International Journal of Nanomedicine 8: 3963.

Kuempel, E., Castranova, V., Geraci, C. and Schulte, P. 2012. Development of risk-based nanomaterial groups for occupational exposure control. Journal of Nanoparticle Research 14: 1–15.

Kurantowicz, N., Strojny, B., Sawosz, E. et al. 2015. Biodistribution of a high dose of diamond, graphite, and graphene oxide nanoparticles after multiple intraperitoneal injections in rats. Nanoscale Research Letters 10(1): 1–14.

Lam, C-w., James, J.T., McCluskey, R. et al. 2006. A review of carbon nanotube toxicity and assessment of potential occupational and environmental health risks. Critical Reviews in Toxicology 36(3): 189–217.

Landsiedel, R., Sauer, U.G. and de Jong, W.H. 2017. Chapter 8 - Risk Assessment and Risk Management. pp. 189–222. *In*: Fadeel, B., Pietroiusti, A. and Shvedova, A.A. (eds.). Adverse Effects of Engineered Nanomaterials (Second Edition). Academic Press.

Lee, D.-K., Jeon, S., Han, Y. et al. 2018. Threshold rigidity values for the asbestos-like pathogenicity of high-aspect-ratio carbon nanotubes in a mouse pleural inflammation model. ACS Nano 12(11): 10867–10879.

Lee, Y.-S., Sung, J.-H., Song, K.-S. et al. 2019. Derivation of occupational exposure limits for multi-walled carbon nanotubes and graphene using subchronic inhalation toxicity data and a multi-path particle dosimetry model. Toxicology Research 8(4): 580–586.

Li, Q., Wang, K., Zheng, Y. et al. 2016. Explosion severity of micro-sized aluminum dust and its flame propagation properties in 20 L spherical vessel. Powder Technology 301: 1299–1308.

Liang, F. and Chen, B. 2010. A review on biomedical applications of single-walled carbon nanotubes. Current Medicinal Chemistry 17(1): 10–24.

Lin, S., Keskar, G., Wu, Y. et al. 2006. Detection of phospholipid-carbon nanotube translocation using fluorescence energy transfer. Applied Physics Letters 89(14): 143118.

Llaver, M., Fiorentini, E.F., Oviedo, M.N., Quintas, P.Y. and Wuilloud, R.G. 2021. Elemental speciation analysis in environmental studies: Latest trends and ecological impact. International Journal of Environmental Research and Public Health 18(22): 12135.

Long, X., Yan, J., Zhang, Z. et al. 2022. Autophagy-targeted nanoparticles for effective cancer treatment: Advances and outlook. NPG Asia Materials 14(1): 1–16.

Lopez-Chaves, C., Soto-Alvaredo, J., Montes-Bayon, M. et al. 2018. Gold nanoparticles: Distribution, bioaccumulation and toxicity. *In vitro* and *in vivo* studies. Nanomedicine: Nanotechnology, Biology and Medicine 14(1): 1–12.

MacCorkle, R., Slattery, S., Nash, D. et al. 2006. Intracellular protein binding to asbestos induces aneuploidy in human lung fibroblasts. Cell Motility and the Cytoskeleton 63(10): 646–657.

Madannejad, R., Shoaie, N., Jahanpeyma, F. et al. 2019. Toxicity of carbon-based nanomaterials: Reviewing recent reports in medical and biological systems. Chemico-biological Interactions 307: 206–222.

Magdolenova, Z., Drlickova, M., Henjum, K. et al. 2015. Coating-dependent induction of cytotoxicity and genotoxicity of iron oxide nanoparticles. Nanotoxicology 9(sup1): 44–56.

Malhotra, N., Ger, T.-R., Castillo, A.L. et al. 2020. An Update Report on the Biosafety and Potential Toxicity of Fullerene-Based Nanomaterials toward Aquatic Animals.

Maynard, A. and Michelson, E. 2006. The nanotechnology consumer products inventory. Woodrow Wilson International Center for Scholars, Washington, DC, accessed March 23.

McCullen, S.D., Stevens, D.R., Roberts, W.A. et al. 2007. Characterization of electrospun nanocomposite scaffolds and biocompatibility with adipose-derived human mesenchymal stem cells. International Journal of Nanomedicine 2(2): 253–263.

Mohammadian, Y. and Nasirzadeh, N. 2021. Toxicity risks of occupational exposure in 3D printing and bioprinting industries: A systematic review. Toxicology and Industrial Health 37(9): 573–584.

Mohammadian, Y., Rezazadeh Azari, M., Peirovi, H. et al. 2019. Combined toxicity of multi-walled carbon nanotubes and benzo [a] pyrene in human epithelial lung cells. Toxin Reviews 38(3): 212–222.

Morimoto, Y., Horie, M., Kobayashi, N. et al. 2013. Inhalation toxicity assessment of carbon-based nanoparticles. Accounts of Chemical Research 46(3): 770–781.

Mukherjee, S., Sau, S., Madhuri, D. et al. 2016. Green synthesis and characterization of monodispersed gold nanoparticles: Toxicity study, delivery of doxorubicin and its bio-distribution in mouse model. Journal of Biomedical Nanotechnology 12(1): 165–181.

Mukhopadhyay, T., Mahata, A., Adhikari, S. et al. 2017. Effective elastic properties of two dimensional multiplanar hexagonal nanostructures. 2D Materials 4(2): 025006.

Murray, A., Kisin, E., Leonard, S. et al. 2009. Oxidative stress and inflammatory response in dermal toxicity of single-walled carbon nanotubes. Toxicology 257(3): 161–171.

Murugadoss, S., Lison, D., Godderis, L. et al. 2017. Toxicology of silica nanoparticles: An update. Archives of Toxicology 91(9): 2967–3010.

Myojo, T. and Ono-Ogasawara, M. 2018. Review; Risk assessment of aerosolized SWCNTs, MWCNTs, fullerenes and carbon black. KONA Powder and Particle Journal 35: 80–88.

Nag, A., Alahi, M.E.E., Mukhopadhyay, S.C. et al. 2021. Multi-walled carbon nanotubes-based sensors for strain sensing applications. Sensors 21(4): 1261.

Nagai, H. and Toyokuni, S. 2010. Biopersistent fiber-induced inflammation and carcinogenesis: Lessons learned from asbestos toward safety of fibrous nanomaterials. Archives of Biochemistry and Biophysics 502(1): 1–7.

Napierska, D., Thomassen, L.C., Lison, D. et al. 2010. The nanosilica hazard: Another variable entity. Particle and Fibre Toxicology 7(1): 1–32.

Nasirzadeh, N., Azari, M.R., Rasoulzadeh, Y. and Mohammadian, Y. 2019. An assessment of the cytotoxic effects of graphene nanoparticles on the epithelial cells of the human lung. Toxicology and Industrial Health 35(1): 79–87.

Nasirzadeh, N., Rasoulzadeh, Y., Rezazadeh Azari, M. et al. 2020. Cellular toxicity of multi-walled carbon nanotubes on human lung cells. Journal of Chemical Health Risks 10(2): 135–144.

Neltner, B.B.T. 2010. Hybrid bio-templated catalysts. Massachusetts Institute of Technology.

Neto, A.C., Guinea, F. and Peres, N.M. 2006. Drawing conclusions from graphene. Physics World 19(11): 33.

Ng, C.T., Yong, L.Q., Hande, M.P. et al. 2017. Zinc oxide nanoparticles exhibit cytotoxicity and genotoxicity through oxidative stress responses in human lung fibroblasts and Drosophila melanogaster. International Journal of Nanomedicine 12: 1621.

Nowack, B. and Bucheli, T.D. 2007. Occurrence, behavior and effects of nanoparticles in the environment. Environmental Pollution 150(1): 5–22.

Nthwane, Y.B., Tancu, Y., Maity, A. and Thwala, M. 2019. Characterisation of titanium oxide nanomaterials in sunscreens obtained by extraction and release exposure scenarios. SN Applied Sciences 1: 1–10.

Oberdörster, G., Maynard, A., Donaldson, K., Castranova, V., Fitzpatrick, J., Ausman, K. et al. 2005. Principles for characterizing the potential human health effects from exposure to nanomaterials: elements of a screening strategy. Particle and Fibre Toxicology 2(1): 1–35.

Omari Shekaftik, S. and Nasirzadeh, N. 2021. 8-Hydroxy-2'-deoxyguanosine (8-OHdG) as a biomarker of oxidative DNA damage induced by occupational exposure to nanomaterials: a systematic review. Nanotoxicology 15(6): 850–864. https://doi.org/10 .1080/17435390.2021.1936254.

Omari Shekaftik, S.H., Shirazi, F., Yarahmadi, R. et al. 2022a. Investigating the relationship between occupational exposure to nanomaterials and symptoms of nanotechnology companies' employees. Archives of Environmental & Occupational Health 77(3): 209–218.

Omari Shekaftik, S. and Nasirzadeh, N. 2021. 8-Hydroxy-2'-deoxyguanosine (8-OHdG) as a biomarker of oxidative DNA damage induced by occupational exposure to nanomaterials: A systematic review. Nanotoxicology 15(6): 850–864.

Omari Shekaftik, S., Nasirzadeh, N., Baba-Ahangar, T. et al. 2022b. Academic nanotechnology laboratories: Investigating good practices and students' health status. Journal of Nanoparticle Research 24(9): 1–16.

Omari Shekaftik, S., Yarahmadi, R., Moghadasi, N. et al. 2020. Investigation of recommended good practices to reduce exposure to nanomaterials in nanotechnology laboratories in Tehran, Iran. Journal of Nanoparticle Research 22(3): 1–8.

Ong, L.-C., Chung, F.F.-L., Tan, Y.-F. et al. 2016. Toxicity of single-walled carbon nanotubes. Archives of Toxicology 90(1): 103–118.

Oroojalian, F., Beygi, M., Baradaran, B. et al. 2021. Immune cell membrane-coated biomimetic nanoparticles for targeted cancer therapy. Small 17(12): 2006484.

Oshimura, M., Hesterberg, T.W., Tsutsui, T. et al. 1984. Correlation of asbestos-induced cytogenetic effects with cell transformation of Syrian hamster embryo cells in culture. Cancer Research 44(11): 5017–5022.

Ou, L., Song, B., Liang, H. et al. 2016. Toxicity of graphene-family nanoparticles: A general review of the origins and mechanisms. Particle and Fibre Toxicology 13(1): 1–24.

Pakrashi, S., Dalai, S., Sabat, D. et al. 2011. Cytotoxicity of Al2O3 nanoparticles at low exposure levels to a freshwater bacterial isolate. Chemical Research in Toxicology 24(11): 1899–1904.

Park, E.-J., Khaliullin, T.O., Shurin, M.R. et al. 2018. Fibrous nanocellulose, crystalline nanocellulose, carbon nanotubes, and crocidolite asbestos elicit disparate immune responses upon pharyngeal aspiration in mice. Journal of Immunotoxicology 15(1): 12–23.

Patel, J. and Patel, A. 2015. Toxicity of nanomaterials on the liver, kidney, and spleen. Biointeractions of Nanomaterials 1: 286–306.

Pattan, G. and Kaul, G. 2014. Health hazards associated with nanomaterials. Toxicology and Industrial Health 30(6): 499–519.

Petersen, E.J., Reipa, V., Watson, S.S. et al. 2014. DNA damaging potential of photoactivated P25 titanium dioxide nanoparticles. Chemical Research in Toxicology 27(10): 1877–1884.

Pfaff, M. 2022. Occupational Safety and Ship Safety, Fire Protection. Ship Operation Technology. Springer, pp. 347–403.

Pietroiusti, A., Iavicoli, I., Bergamaschi, A. et al. 2009. Toxic effects of single-walled carbon nanotubes on the cardiovascular system: State of art. International Journal of Environment and Health 3(3): 264–274.

Poborilova, Z., Opatrilova, R. and Babula, P. 2013. Toxicity of aluminium oxide nanoparticles demonstrated using a BY-2 plant cell suspension culture model. Environmental and Experimental Botany 91: 1–11.

Poland, C.A., Duffin, R., Kinloch, I. et al. 2008. Carbon nanotubes introduced into the abdominal cavity of mice show asbestos-like pathogenicity in a pilot study. Nature Nanotechnology 3(7): 423–428.

Poulsen, S.S., Saber, A.T., Mortensen, A. et al. 2015. Changes in cholesterol homeostasis and acute phase response link pulmonary exposure to multi-walled carbon nanotubes to risk of cardiovascular disease. Toxicology and Applied Pharmacology 283(3): 210–222.

Pritchard, D. 2004. Literature review: Explosion hazards associated with nanopowders. Health and Safety Laboratory Buxton, England.

Proquin, H., Rodríguez-Ibarra, C., Moonen, C.G. et al. 2017. Titanium dioxide food additive (E171) induces ROS formation and genotoxicity: contribution of micro and nano-sized fractions. Mutagenesis 32(1): 139–149.

Roberts, A.P., Mount, A.S., Seda, B. et al. 2007. *In vivo* biomodification of lipid-coated carbon nanotubes by Daphnia magna. Environ. Sci. Technol. 41(8): 3025–3029.

Ryu, H.J., Seong, N.-w., So, B.J. et al. 2014. Evaluation of silica nanoparticle toxicity after topical exposure for 90 days. International Journal of Nanomedicine 9(Suppl 2): 127.

Sadeghi, L., Tanwir, F. and Babadi, V.Y. 2015. *In vitro* toxicity of iron oxide nanoparticle: Oxidative damages on Hep G2 cells. Experimental and Toxicologic Pathology 67(2): 197–203.

Sakamoto, Y., Nakae, D., Fukumori, N. et al. 2009. Induction of mesothelioma by a single intrascrotal administration of multi-wall carbon nanotube in intact male Fischer 344 rats. The Journal of Toxicological Sciences 34(1): 65–76.

Samadi, A., Klingberg, H., Jauffred, L. et al. 2018) Platinum nanoparticles: A non-toxic, effective and thermally stable alternative plasmonic material for cancer therapy and bioengineering. Nanoscale 10(19): 9097–9107.

Sargent, L., Hubbs, A., Young, S.-H. et al. 2012. Single-walled carbon nanotube-induced mitotic disruption. Mutation Research/Genetic Toxicology and Environmental Mutagenesis 745(1-2): 28–37.

Savolainen, K., Pylkkänen, L., Norppa, H., Falck, G., Lindberg, H., Tuomi, T. et al. 2010. Nanotechnologies, engineered nanomaterials and occupational health and safety—A review. Safety Science 48(8): 957–963.

Scrivens, W.A., Tour, J.M., Creek, K.E. et al. 1994. Synthesis of 14C-labeled C60, its suspension in water, and its uptake by human keratinocytes. Journal of the American Chemical Society 116(10): 4517–4518.

Seabra, A.B., Paula, A.J., de Lima, R. et al. 2014. Nanotoxicity of graphene and graphene oxide. Chemical Research in Toxicology 27(2): 159–168.

Seiffert, S.B., Vennemann, A., Nordhorn, I.D. et al. 2022. LA-ICP-MS and immunohistochemical staining with lanthanide-labeled antibodies to study the uptake of CeO2 nanoparticles by macrophages in tissue sections. Chemical Research in Toxicology.

Senapati, V.A. and Kumar, A. 2018. ZnO nanoparticles dissolution, penetration and toxicity in human epidermal cells. Influence of pH. Environmental Chemistry Letters 16(3): 1129–1135.

Sengul, A.B. and Asmatulu, E. 2020. Toxicity of metal and metal oxide nanoparticles: A review. Environmental Chemistry Letters 18(5): 1659–1683.

Sharma, A., Hosseini-Bandegharaei, A., Kumar, N. et al. 2022. Insight into ZnO/carbon hybrid materials for photocatalytic reduction of CO2: An in-depth review. Journal of CO2 Utilization 65: 102205.

Sharma, M., Nikota, J., Halappanavar, S. et al. 2016. Predicting pulmonary fibrosis in humans after exposure to multi-walled carbon nanotubes (MWCNTs). Archives of Toxicology 90(7): 1605–1622.

Sharma, N., Gupta, R.D., Sharma, R.C. et al. 2021. Graphene: An overview of its characteristics and applications. Materials Today: Proceedings 47: 2752–2755.

Shipelin, V., Smirnova, T., Gmoshinskii, I. et al. 2015. Analysis of toxicity biomarkers of fullerene C60 nanoparticles by confocal fluorescent microscopy. Bulletin of Experimental Biology and Medicine 158(4): 443–449.

Singh, S.K., Singh, M.K., Nayak, M.K. et al. 2011. Thrombus inducing property of atomically thin graphene oxide sheets. ACS Nano 5(6): 4987–4996.

Srikanth, K., Mahajan, A., Pereira, E. et al. 2015. Aluminium oxide nanoparticles induced morphological changes, cytotoxicity and oxidative stress in Chinook salmon (CHSE-214) cells. Journal of Applied Toxicology 35(10): 1133–1140.

Steckiewicz, K.P., Barcinska, E., Malankowska, A. et al. 2019. Impact of gold nanoparticles shape on their cytotoxicity against human osteoblast and osteosarcoma in *in vitro* model. Evaluation of the safety of use and anti-cancer potential. Journal of Materials Science: Materials in Medicine 30(2): 1–15.

Subramaniam, V.D., Prasad, S.V., Banerjee, A. et al. 2019. Health hazards of nanoparticles: understanding the toxicity mechanism of nanosized ZnO in cosmetic products. Drug and Chemical Toxicology 42(1): 84–93.

Tan, K.X., Barhoum, A., Pan, S. et al. 2018. Risks and toxicity of nanoparticles and nanostructured materials. Emerging applications of nanoparticles and architecture nanostructures. Elsevier, pp. 121–139.

Thota, S. and Crans, D.C. 2018. Metal Nanoparticles: Synthesis and Applications in Pharmaceutical Sciences. John Wiley and Sons.

Ursini, C.L., Cavallo, D., Fresegna, A.M. et al. 2014. Evaluation of cytotoxic, genotoxic and inflammatory response in human alveolar and bronchial epithelial cells exposed to titanium dioxide nanoparticles. Journal of Applied Toxicology 34(11): 1209–1219.

Ursini, C.L., Fresegna, A.M., Ciervo, A. et al. 2021. Occupational exposure to graphene and silica nanoparticles. Part II: Pilot study to identify a panel of sensitive biomarkers of genotoxic, oxidative and inflammatory effects on suitable biological matrices. Nanotoxicology 15(2): 223–237.

Vallyathan, V., Mega, J.F., Shi, X. et al. 1992. Enhanced generation of free radicals from phagocytes induced by mineral dusts. Am. J. Respir. Cell Mol. Biol. 6(4): 404–413.

Viegas, S. 2020. The Relevance of Hand-Mouth Contact in Occupational Exposure to Metals. Occupational and Environmental Safety and Health II. Springer, pp. 359–365.

Vileno, B., Sienkiewicz, A., Lekka, M. et al. 2004. *In vitro* assay of singlet oxygen generation in the presence of water-soluble derivatives of C60. Carbon 42(5-6): 1195–1198.

Wan, R., Mo, Y., Zhang, Z. et al. 2017. Cobalt nanoparticles induce lung injury, DNA damage and mutations in mice. Particle and Fibre Toxicology 14(1): 1–15.

Wang, H., Ma, Y., Shen, Y. et al. 2022. Designing of Co3O4/N-doped carbon and Co-Co3O4/N-doped carbon nanomaterials via sol-gel route with unexpected high catalytic performances toward hydrogenation reduction of 4-Nitrophenol. Journal of Alloys and Compounds 923: 166403.

Wang, S., Cole, I.S. and Li, Q. 2016. The toxicity of graphene quantum dots. RSC Advances 6(92): 89867–89878.

Wang, X., Dong, A., Hu, Y. et al. 2020. A review of recent work on using metal–organic frameworks to grow carbon nanotubes. Chemical Communications 56(74): 10809–10823.

Wang, Y., Liu, J., Liu, L. et al. 2011. High-quality reduced graphene oxide-nanocrystalline platinum hybrid materials prepared by simultaneous co-reduction of graphene oxide and chloroplatinic acid. Nanoscale Research Letters 6(1): 1–8.

Wen, K.P., Chen, Y.C., Chuang, C.H. et al. 2015. Accumulation and toxicity of intravenously-injected functionalized graphene oxide in mice. Journal of Applied Toxicology 35(10): 1211–1218.

Wu, K., Zhou, Q. and Ouyang, S. 2021. Direct and indirect genotoxicity of graphene family nanomaterials on DNA—A review. Nanomaterials 11(11): 2889.

Xiang, H., Feng, W. and Chen, Y. 2020. Single-atom catalysts in catalytic biomedicine. Advanced Materials 32(8): 1905994.

Xiang, K., Dou, Z., Li, Y. et al. 2012. Cytotoxicity and Tnf-A secretion in Raw264. 7 macrophages exposed to different fullerene derivatives. Journal of Nanoscience and Nanotechnology 12(3): 2169–2178.

Xiao, L., Liu, C., Chen, X. et al. 2016. Zinc oxide nanoparticles induce renal toxicity through reactive oxygen species. Food and Chemical Toxicology 90: 76–83.

Xin, X., Huang, G., An, C. et al. 2019. Interactive toxicity of triclosan and nano-TiO_2 to green alga Eremosphaera viridis in Lake Erie: A new perspective based on Fourier transform infrared spectromicroscopy and synchrotron-based X-ray fluorescence imaging. Environmental Science & Technology 53(16): 9884–9894.

Yah, C.S., Iyuke, S.E. and Simate, G.S. 2012. A review of nanoparticles toxicity and their routes of exposures. Iranian Journal of Pharmaceutical Sciences 8(1): 299–314.

Yan, L., Zhao, F., Li, S. et al. 2011. Low-toxic and safe nanomaterials by surface-chemical design, carbon nanotubes, fullerenes, metallofullerenes, and graphenes. Nanoscale 3(2): 362–382.

Yang, K., Gong, H., Shi, X. et al. 2013. *In vivo* biodistribution and toxicology of functionalized nano-graphene oxide in mice after oral and intraperitoneal administration. Biomaterials 34(11): 2787–2795.

Yazdimamaghani, M., Moos, P.J., Dobrovolskaia, M.A. et al. 2019. Genotoxicity of amorphous silica nanoparticles: Status and prospects. Nanomedicine: Nanotechnology, Biology and Medicine 16: 106–125.

Young, T.L., Mostovenko, E., Denson, J.L. et al. 2021. Pulmonary delivery of the broad-spectrum matrix metalloproteinase inhibitor marimastat diminishes multiwalled carbon nanotube-induced circulating bioactivity without reducing pulmonary inflammation. Particle and Fibre Toxicology 18(1): 1–16.

Zhang, S., Yi, K., Chen, A., Shao, J., Peng, L. and Luo, S. 2022. Toxicity of zero-valent iron nanoparticles to soil organisms and the associated defense mechanisms: a review. Ecotoxicology, 1–11.

Zhang, T., Tang, M., Zhang, S. et al. 2017. Systemic and immunotoxicity of pristine and PEGylated multi-walled carbon nanotubes in an intravenous 28 days repeated dose toxicity study. International Journal of Nanomedicine 12: 1539.

Zhang, X., Yin, J., Peng, C. et al. 2011. Distribution and biocompatibility studies of graphene oxide in mice after intravenous administration. Carbon 49(3): 986–995.

CHAPTER 8

Nanotechnology Innovation and its Environmental Effects

Iqra Noshair, Shifa Rabani, Farkhanda Manzoor
and *Zakia Kanwal**

1. Introduction

Nanotechnology is known as a general-purpose technology due to the fact that, in its advanced form, it will have a significant impact on almost every industry and all the aspects of society. It will provide better-built, longer-lasting, cleaner, safer and smarter products for homes, communication, medicine, transportation, agriculture, and the general industry (Li et al. 2010).

The distinct physical characteristics of materials at the nanoscale have drawn a lot of attention to nanotechnology in recent years. Due to their higher surface-to-volume ratios compared to their bulkier counterparts, nanomaterials exhibit enhanced reactivity and, in turn, remarkable effectiveness. In addition, compared to conventional methods, nanomaterials have the potential to take advantage of special surface chemistry, enabling them to be functionalized or grafted with functional groups that can target particular molecules of interest (pollutants) for effective remediation. Additionally, the deliberate tinkering with the physical characteristics of nanomaterials, such as their morphology, chemical composition, size and permeability, can impart additional beneficial qualities that directly influence how well these materials perform for contaminant remediation (Campbell et al. 2015). The tunable physical parameters and rich surface modification chemistry of nanomaterials have significant advantages over traditional methods for addressing environmental contamination.

Department of Zoology, Faculty of Natural Sciences, Lahore College for Women University, Jail Road Lahore, 54000, Pakistan.
* Corresponding author: zakia.kanwal@lcwu.edu.pk

As a result, techniques that are created as a blend of several different materials (hybrids/composites), obtaining particular desired properties from each of their constituents, may be more effective, specific and stable than techniques built upon a single nanoplatform. As an alternative to using nanoparticles (NPs) alone, NPs adhering to a scaffold may be able to increase the stability of the material. The selectivity and effectiveness of a material can be improved by functionalizing it with particular chemicals that target desired contaminants (Guerra et al. 2017).

Furthermore, the nanotechnology revolution is characterized by its interdisciplinary nature, bringing together scientists from various fields of study. As a result, there has been a significant improvement in the quality of machinery, automobiles, electronics, common household appliances, consumer products, agriculture, microscopy, scientific instruments, healthcare, diagnostic assays and drug discovery (DiSia 2017). The global collaboration among researchers has contributed to the widespread applications of nanotechnology in diverse sectors, making it a transformative force shaping our modern world (Figure 1).

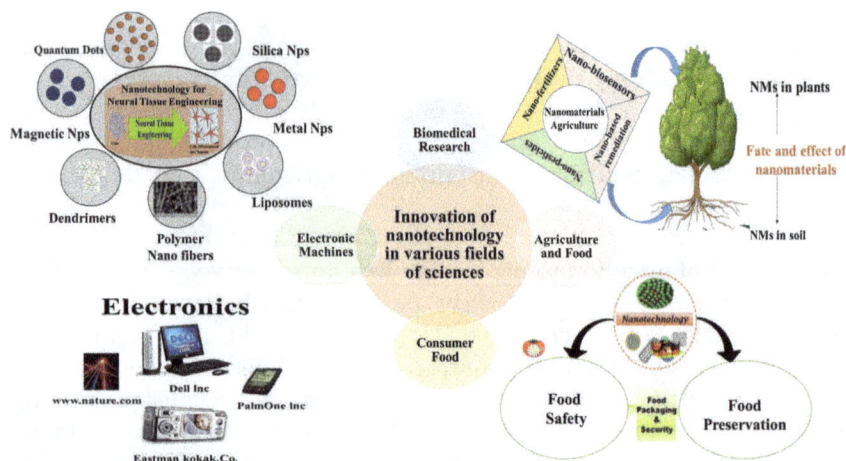

Figure 1: Different nanomaterials used for agriculture, biomedical research, food and electronics: nanotechnology and their route of applications.

2. Innovation of Nanotechnology in Various Fields of Science

2.1 Membranes and Membrane Processes

Applications of membrane separation techniques for the treatment of water and wastewater are developing quickly. Depending on the sizes of the pores and the molecules, membranes act as a physical barrier for various substances. In the water and wastewater industry, membrane technology is recognized as a dependable and largely automated process. By creating antifouling layers, for example, new nanotechnology research activities are concentrating on enhancing selectivity and flux efficiency. The state of the art for nanoengineered membrane filtration is covered in the following sections.

2.2 Nanofiltration Membranes

One method of membrane filtration, known as nanofiltration is a pressure-driven procedure where molecules and particles with a diameter of 0.5 nm to 1 nm or less are rejected by the membrane. A special charge-based repulsion mechanism that enables the separation of different ions is what distinguishes nanofiltration membranes from other membranes (Sharma and Sharma 2012, Jagadevan et al. 2012, Qu et al. 2013). They are primarily used to reduce groundwater hardness, colour, odour and the concentration of heavy metal ions therein. Desalination, the process of turning seawater into potable water, is another advantageous area of application as the competing desalination technologies are very expensive (Table 1).

Table 1: The most significant characteristics, uses and novel strategies of membranes and membrane processes for water and wastewater treatment processes.

Nano membranes	Properties		Applications	Novel Approaches
	Positive	Negative		
Nanofiltration membrane	Charge-base, repulsion, high selectivity, relatively low pressure	Membrane blocking	Applied for the reduction of hardness, balancing of heavy metal ions and for color removing from groundwater	Sea water desalination
Nanocomposite membrane	Increased hydrophilicity of water, fouling resistance, water permeability, thermal and chemical robustness	Resistance of bulk material when using oxidizing nanomaterials, possible release of NPs	Used for osmosis reversal and removal of micropollutants	Bio nanocomposite membrane

2.3 Nanocomposite Membranes

Mixed matrix membranes and surface-functionalized membranes make up the new class of filtration materials known as nanocomposite membranes. Herein, nanofillers are used in mixed matrix membranes and are incorporated into the matrix material; these nanofillers are typically inorganic and are incorporated into a polymeric or inorganic oxide matrix; they also have a higher surface-to-mass ratio due to their greater specific surface area (Wegmann et al. 2008, Feng et al. 2013). Al_2O_3, TiO_2 and other metal oxide NPs can help improve the mechanical and thermal stability as well as the permeate flux of polymeric membranes. Zeolites are incorporated into membranes to increase their hydrophilicity, which raises their water permeability. The main methods for boosting fouling resistance involve the use of antimicrobial NPs(such as nanosilver and CNTs) and (photo) catalytic nanomaterials (such as bimetallic NPs and TiO_2).

2.4 Innovation of Nanotechnology in Medicine

The four primary sub-disciplines that make up the developing field of nanomedicine are drug delivery systems, nanodevices, chemotherapy analytical tools and nanoimaging. One could say that due to the potential benefits of nanobiotechnology, there are high hopes for treatment of cancer, viruses and fungi, diabetes, chronic lung diseases, and gene therapy. In addition to medical therapy, nanomedicine can also be used in surgery, in photodynamic therapy for instance (Bawa et al. 2016). Following are some applications of nanotechnology in medicine.

2.5 Nanomedicine for Diagnosis

Nanotechnology can help doctors make diagnoses at the level of a single cell or molecule (Hulla et al. 2015). The most widely used NPs are quantum dots, silver NPs and gold NPs. While various other nanotechnological tools for instance, nanobiosensors are available for potential clinical applications, their development and integration into medical practice requires thorough testing, safety assessment, and regulatory approval to ensure their efficacy and safety in patient care. Nanotechnology will also help advance point-of-care diagnostics, the combination of diagnostics and therapeutics, and personalized medicine, by pushing the boundaries of current molecular diagnostics. At present, the most significant applications being exercised in diagnostics are biomarker discovery, biochips, early cancer diagnosis and detection of infectious germs, all showing that the potential diagnostics used are not limited (Jain 2004).

2.6 Medicine and Healthcare

The application of nanotechnology may alter how we approach some of the most pressing issues in global development. In 2005, The Millennium Project Taskforce on Science, Technology, and Innovation of the United Nations (UN) arrived at the conclusion that nanotechnology can help achieve the Millennium Development Goals (MDGs), specifically the targets to lower child mortality, raise maternal mortality, and combat HIV/AIDS, malaria and other diseases (Freitas 2005).

2.7 Disease Screening and Diagnosis

Numerous options for disease detection, diagnosis, monitoring and treatment are offered by nanotechnology. Quantum dots, a type of semiconducting nanocrystal, can, for instance, improve biological imaging for medical diagnosis (Mathuria 2009). They emit a wide spectrum of vivid colours when exposed to ultraviolet light, which can be used to locate and distinguish between different types of cells and biological processes. Compared to traditional dyes which are used in many biological tests like MRIs, these crystals provide up to 1000 times better optical detection and significantly more information (Gardner 2015). Researchers today are thus looking into how nanotechnology might be used to create portable point-of-care diagnostic kits that would allow the simultaneous and accurate identification of multiple diseases (Yen et al. 2015).

2.8 Drug Development and Delivery System

The invention of novel therapeutic approaches and drug delivery systems can also be facilitated by nanotechnology (Gardner 2015). Using nanotechnology, drugs can be more precisely delivered to target body parts through nanovehicles (Mamo et al. 2010). Additionally, the formulation of medications can be changed to improve how well the active component penetrates cell membranes, thus lowering the dosage required (Gardner and Dhai 2014). Therefore, by addressing issues like drug toxicity, maintaining the release of drugs in the body, enhancing bioavailability, and reducing the amount of active ingredients per dose, nanotechnology can revolutionize the delivery of drugs. Patient compliance may increase as a result of its potential to shorten the time required for drug administration and reduce drug side effects, which will ultimately help with the effective management and treatment of the disease (Gardner 2015).

3. Tissue Engineering

Research in fields like toxicogenomics, synthetic biology, regenerative medicine and genetic modification are all advancing with the use of nanotechnology tools and techniques (Musee et al. 2010). For instance, studies are being conducted on the use of nanotechnology to encourage the growth of nerve cells in damaged brain or spinal cord cells for instance. In one technique, a nanostructured gel fills the gaps between existing cells and promotes the growth of new cells. Another investigation is looking into the use of nanofibers to repair damaged spinal nerves (Gardner 2015).

3.1 Agriculture

The application of nanotechnology in agriculture is anticipated to maximize productivity and address the issues in the food sector. Food and adequate nutrition are in high demand, but the world's grain harvest has failed to keep up with the demand over the past four years. With 9.4 hectares of forest being lost each year, biodiversity is being destroyed globally. A quarter of the world's coral reefs and half of our forests have vanished. By the year 2050, the current global population of 6.4 billion people is expected to increase to 8.9 billion. It is predicted that 98% of this growth will occur in underdeveloped nations. By 2030, it is expected that there will be 5 billion people living in cities, putting significant strain on food production and distribution. Precision farming using nanosensors, nanopesticides and low-cost decentralized water purification using nanotechnology would likely solve the problems. Moreover, plant gene therapy will be a more sophisticated nanotechnological solution, resulting in pest-resistant high-yield crops that use less water (Aithal and Aithal 2015).

3.2 Drinking Water

The water sector is one of the most important environmental applications of nanotechnology. Scientists are starting to think about using seawater as a source of drinking water as freshwater sources become increasingly scarce due to overuse and contamination. The majority of the world's water supply contains too much salt for human consumption, and desalination is one possible but expensive way to get rid of the salt and create new sources of drinking water. Membranes made of carbon nanotubes could lower the price of desalination. Nanofilters can also be used to remediate or clean up ground water or surface water that has been contaminated with harmful chemicals. Furthermore, waterborne contaminants can be detected using nanosensors (Yashwant et al. 2022).

Nanotechnology has the potential to offer effective, affordable and ecologically sustainable methods of supplying potable water for human consumption as well as clean water for agricultural and industrial uses. The world's problem with access to clean drinking water is expected to be solved by nanotechnological innovations in low-cost water purification. One of the most valuable natural resources on Earth is water, and most of it is saltwater. Only 3% of the world's freshwater supply is usable, and two-thirds of that is frozen in glaciers, ice caps and icebergs; human consumption is possible with the remaining 1% only. The need for fresh water is rising. Currently, 70% of the water in the world is used for agriculture. By 2030, there will be a 60% increase in the demand of water supply to feed an additional 2 billion people. By 2050, about two-thirds of the world's population will be affected by droughts due to current consumption, population and development rates. This problem can be solved by nanotechnology through low-cost decentralized water purification, molecular level contaminant detection and significantly enhanced filtration systems. This facilitates the inexpensive conversion of seawater into drinking water (Nagar and Pradeep 2020).

4. Innovation of Nanotechnology in the Food Industry

Food nanosensing and food nanostructured ingredients are the two main applications of nanotechnology in the food industry (Singh et al. 2013) Nanostructured food ingredients are used in a variety of applications, including food processing and packaging, while food nanosensing improves food quality and safety (Figure 2).

Nanostructures and nanostructured materials can be used in food processing as: (a) food additives and carriers for effective nutrient delivery, (b) anti-caking agents, (c) gelling agents, and (d) nanocapsules and nanocarriers for preserving flavour, aroma, etc.; other food ingredients are just a few examples of the types of food additives and carriers that can enhance food quality nutritionally. Food nano-packaging takes into account enhanced packaging, active packaging, intelligent packaging and bio-based packaging.

4.1 Food Packaging

One of the most important aspects in the food safety process is food packaging. The main goals of food packaging are to avoid contamination and spoiling, boost sensitivity by permitting enzyme activity, and limit weight loss (He et al. 2019). Because of the harsh environment, bioactive components in functional foods are frequently degraded and inactivated, thus reducing the product's shelf life. Use of nanostructured

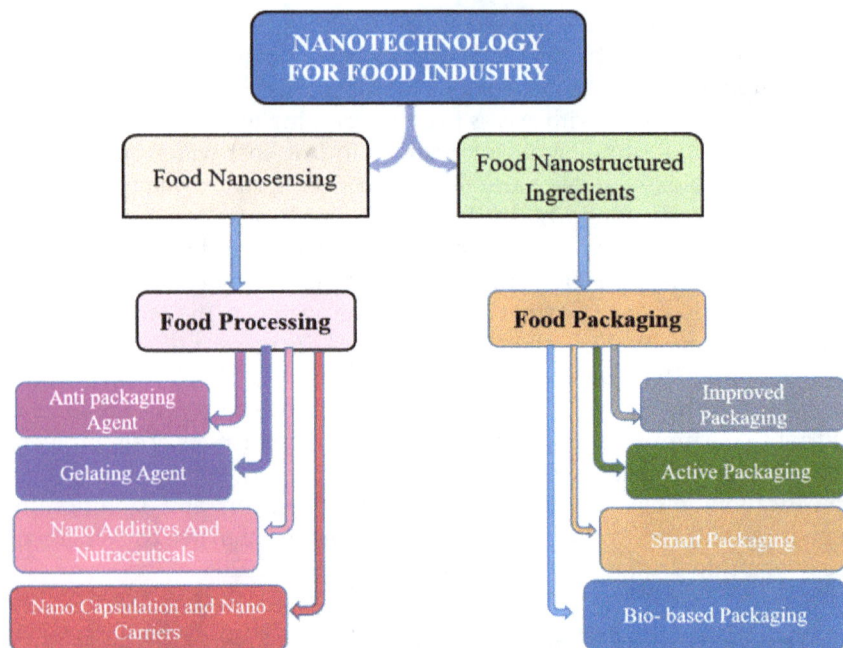

Figure 2: Use of (bio) nanotechnology in different fields of the food industry.

or nanomodified materials is a viable method for extending and ensuring a food product's shelf life (Bajpai et al. 2018).

Food packaging can be improved by using functional nanomaterials with enhanced physico-chemical properties, such as greater mechanical strength, durability and flexibility (improved packaging). Packaging can also be made better by using nanomaterials with active characteristics like antimicrobial, antioxidant and UV protection properties, as well as nanosensors with smart or intelligent qualities like those that can detect gases and small organic molecules, function as active stages, and identify products (smart packaging). Bio-based packaging can benefit from the usage of bionanomaterials, being more biodegradable, biocompatible, waste-free and environmentally friendly (Ranjan et al. 2014). Bio-based packaging, such as biodegradable packaging and biocompatible packaging, has the potential to replace non-biodegradable plastic polymers typically used as packaging materials at present (Kuswandi 2017). Classification of food packaging properties based on functional nanomaterials has been presented in Figure 3.

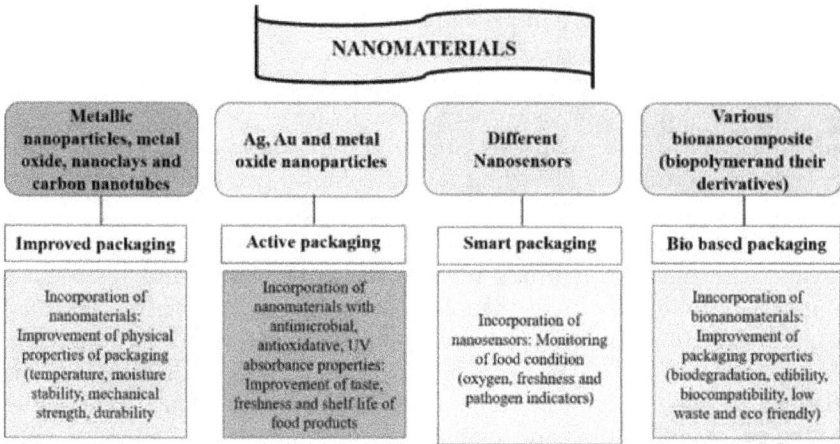

Figure 3: Classification of food packaging based on functional nanomaterials.

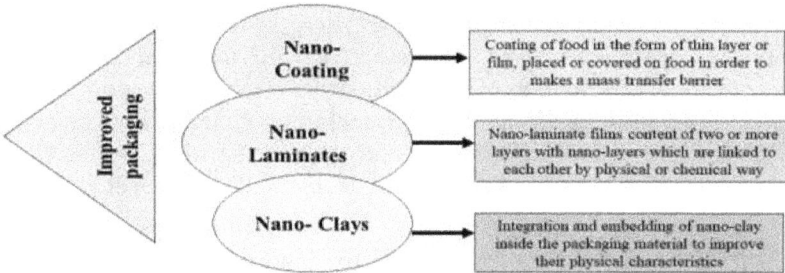

Figure 4: Nanotechnology techniques used for the enhancement of mechanical and physical properties of food packaging.

4.2 Improved Food Packaging

Enhancing the mechanical and physical characteristics of packaging, such as its flexibility, strength, and resistance to temperature and humidity, is the fundamental goal of better food packaging (Figure 4). Functional NPs are integrated into polymer materials to achieve this. For different beverages and oil industries, various nanocomposites or NPs incorporated into polymers and nanostructures are already being produced, with the proportion of NPs (like clay nanoparticle composites) being up to 5% (w/w). These nanomaterials enhance the packaging's barrier properties, reducing oxygen and carbon dioxide permeation by up to 80–90% (Brody 2007). Kim and Cha (2014) reported that excessive clay loadings in ethylene-vinyl alcohol (EVOH) copolymer-based nanocomposite films cause the formation of clay agglomerates, degrading the tensile properties and optical transparency. Compared to the material without montmorillonite clay NPs, the oxygen barrier properties of the nanocomposite films are improved by 59%, with the addition of just 3% (w/w) clay. According to Arora et al. (2011), adding 4% (w/w) micro clay increases the oxygen barrier characteristics of polystyrene (PS) by 51%.

The advantages of using nanomaterials for food packaging, over traditional packaging materials, are numerous. Nano-coating is the most popular nanotechnology technique for enhancing the characteristics of food packaging. The food can be covered with various nanotechnology techniques, such as, food coatings in the form of a thin layer or film, to create a barrier against mass transfer, nano laminates and nano clays (Figure 4). Edible food coatings can be placed directly on the product, unlike non-edible coatings that serve as a protective container (irrespective of whether they are a part of the food product) (Joye et al. 2016). To add flavour, colour, enzymes, antioxidants, and anti-browning compounds to the products, edible nano-coatings (5 nm thin coatings) are applied in the cheese, bakery, agriculture and meat processing industries, among others (Nile et al. 2020, Azeredo et al. 2009). Additionally, due to their mechanical, thermal and barrier properties, as well as their low cost, nano-clays are widely used and researched for food packaging (Gabr et al. 2015). For the creation of nanolaminates for food packaging, a variety of adsorbing substances can be used, including charged lipids, charged polyelectrolytes (such as proteins and polysaccharides), and colloidal particles (such as micelles, vesicles and droplets) (Dasgupta et al. 2016).

5. Radio Frequency Identification: Use of Nanocopper Conductive Ink

A nanocopper conductive ink comprised of copper anti-oxidation NPs can be utilised to improve Radio Frequency Identification (RFID) applications. It is suggested that this innovative method can be used to produce RFID devices more quickly and more affordably, while also expanding the range of possible applications. At first, RFID devices required to be etched, covered in electroplate and were otherwise treated as corrosive (Deng et al. 2012). Now, using conductive ink, RFID devices can be quickly and effectively printed on practically all types of substrates.

6. High–Performance and Miniaturized Power using Nanomagnetics

In order to achieve high-frequency permeability, reduced losses, and enhanced power handling, magnetic structures at the nanoscale are designed Although natural metals like iron and nickel have nanomagnetic properties, their high electrical conductivity makes them unsuitable for high-frequency applications. A copper or iron-based nanocomposite

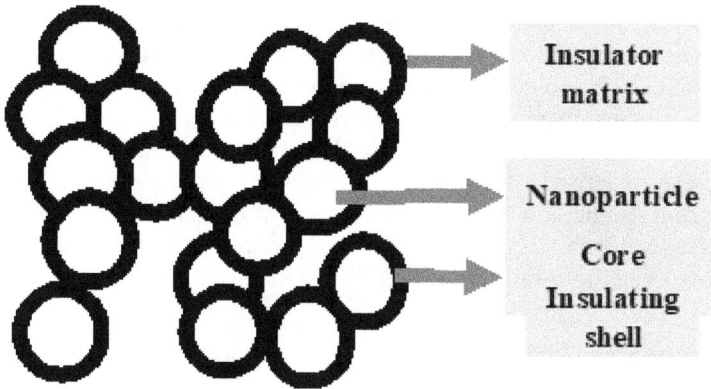

Figure 5: Composite structure of Nanomagnetic materials.

shows higher permeability and frequency stability at microwave frequencies than those obtained from bulk copper or iron metal or their microscale composites as shown in Figure 5. The interaction of magnetic orderings within the molecules, which spreads to the nearby molecules through spin polarization to a distance of 25 nm, is the cause of this phenomena. Due to the interaction's negation of magnetic anisotropy and the interaction's tiny size, little eddy current is produced. Superior magnetic qualities are produced as a result, including high permeability (100–500 H/m), enhanced frequency (1–10 GHz) and strong power management. This is used in power converters and radio frequency inductors, where it is possible to increase volumetric density by up to 10 times, without sacrificing power efficiency. Since antennas have traditionally been huge, it is necessary to surround them with a material that has a high permittivity or permeability in order to reduce their size. This issue can be resolved by using nanomagnetic metal polymers, such as cobalt polymer nanocomposites, which result in increased bandwidth and decreased size. This demonstrates how magnetic materials can be employed at the nanoscale to miniaturize antennas and other radio frequency devices, while also achieving high frequency and better power handling (Raj et al. 2012).

7. Nano-devices in Environmental Sensing Systems

Due to the ease of manufacture, quick detection, high sensitivity and naked-eye sensing of colorimetric sensors, they have demonstrated significant potential for the detection of metallic ions, anions, organic dyes, medicines, pesticides and other harmful contaminants (Aldewachi et al. 2018). Numerous functional nanomaterials (including NPs, quantum dots

(QDs), carbon nanotubes/nanofibers (CNTs/CNFs), nanowires, graphene and other two-dimensional (2D) nanomaterials) have been used to fabricate various sensor and biosensor platforms for environmental and biomedical applications as nanotechnology and materials science have advanced. For instance, surface-modified glassy carbon electrodes promote their use as electrochemical sensors for label-free detection; fluorescent NPs and QDs in fluorescent sensor platforms enable high-performance sensing through fluorescent quenching; and the development of metal NPs-based active substrates can facilitate surface-enhanced Raman scattering determination at as low as a single-molecule level (Liu et al. 2020).

8. Renewable Energy

Renewable energy innovations based on nanotechnology provide all the energy needed by people to meet their basic needs and live comfortably. It is extremely difficult to balance our planet's environmental costs with humanity's need for energy. By 2025, it is expected that global energy demand will have increased by 50%, with fossil fuels accounting for the majority of this increase. Currently, 2.4 billion people rely on plant life, vegetation, or agricultural waste for energy and heating, and over 1.6 billion people do not have access to electricity. By the year 2025, our consumption of fossil fuels may have doubled. Earth's glaciers are melting, atmospheric CO_2 levels have nearly doubled, and three of the last five years have seen the highest temperatures ever recorded since records began in 1861. 1998 was the hottest year on record, 2001 was the second hottest year and 2004 was the hottest year overall. By creating more efficient lighting, fuel cells, hydrogen storage, solar cells, locally distributed power generation, and decentralized generation and storage, nanotechnology can help meet our need for energy solutions (Aithal and Aithal 2016).

9. Air Pollution

Another potential application for nanotechnology is in the reduction of air pollution. To purify indoor air volumes in buildings, filtration techniques comparable to the water purification techniques mentioned above can be used. Nanofilters can be used to separate contaminants and keep them from escaping into the atmosphere through factory smokestacks and automobile tailpipes. Finally, using nanosensors, toxic gas leaks can be detected at incredibly low concentrations. Overall, nanotechnology has a wide range of promising environmental applications. Most of the recent research in nanotechnology has been directed towards water technologies and energy (Mansoori et al. 2008).

Technology is the creation, modification, use and knowledge of tools, machines, techniques, crafts, systems, and methods of organizing them in order to solve problems, improve preexisting solutions to problems, achieve goals, handle applied input or output relations, or perform a particular function. The ability of humans and other animal species to manage and adapt to their natural environments is significantly impacted by technologies. The impact of technology on the society and the environment has been enormous. The ideas of ideal gas, ideal engine, ideal switch, ideal fuel, and ideal semiconductor devices like ideal diodes, transistors and amplifiers have been created and adopted as standards in order to improve the quality and efficacy of such practical equipment or systems. Researchers have determined that the traits and properties of authentic equipment and systems have been regularly modified to boost their performance. Therefore, the optimum characteristics of a system or equipment can be updated or enhanced to obtain 100% efficiency (Aithal and Aithal 2015).

9.1 Pollution Control

Environmental pollution is indubitably one of the biggest issues that society is currently facing. For the removal of contaminants from the soil, water and air, new technologies are being constantly investigated (Masciangoli and Zhang 2003). Examples of some of the numerous alarming contaminants include organic compounds, particulate matter, herbicides, oil spills, fertilizers, toxic gases, sewage, industrial effluents, pesticides and heavy metals (Vaseashta et al. 2007). Since different materials can be used in environmental remediation, a wide range of strategies can be used to achieve this goal. Due to the complexity of the mixing of diverse compounds, and low reactivity and high volatility of environmental pollutants, recent studies have focused on the use of nanomaterials for the development of new environmental remediation methods (Tratnyek and Johnson 2006).

It is essential that the materials used to clean up pollution do not become new sources of pollution after they have been used. Biodegradable materials are therefore very intriguing for this application area. The use of biodegradable materials may boost consumer trust and acceptance of a certain technology because there would not be any material waste left over after treatment, and it may also provide a safer and more environmentally friendly option for the clean-up of pollutants in the environment. In addition, innovative technologies that rely on the target-specific collection of pollutants and can overcome the low efficiency brought on by off-targeting are very alluring. Numerous studies have concentrated on implementing the principles of nanotechnology and integrating them with chemical and physical modification of the surface of materials in order to produce engineered materials that can overcome many of the challenges connected with the remediation of contaminants (Pandey and Fulekar 2012).

One of the most notable examples of a rapidly developing technology with significant potential benefits is the remediation of contaminated groundwater using NPs containing zero-valent iron. This is just one of the many applications of nanotechnology that have an impact on the environment. However, there are a lot of unanswered questions about the core components of this technology, which makes it challenging to engineer applications for their best performance or to evaluate their risk to human or ecological health. Considerations must be given to this key element of NPs as well (Mansoori et al. 2008).

The process of reducing the risks that chemical and radiological contaminants pose to humans and the environment, through sequestration, degradation or other related methods, is known as environmental remediation. The advantages of using nanomaterials for remediation include quicker or more affordable waste cleanup (Kootenaei et al. 2013).

A significant obstacle in the development of effective remediation methods that safeguard the environment is the need for cost-effective remediation techniques. In the case of remediation of soils, sediments and groundwater, heavy metals (such as lead, cadmium and mercury) and organic compounds (e.g., creosote, chlorinated solvents, toluene and benzene) are among the substances of greatest concern. Increased selectivity, affinity and capacity for pollutants may be achieved through precise molecular control and design of materials. Among the objectives of environmental protection agencies are the reduction of hazardous waste and their exposure to air and water, as well as the provision of clean drinking water. In this regard, nanotechnology might be a crucial component of technologies for preventing pollution (Li et al. 2019).

The development of new nanomaterials for environmental remediation presents several important challenges that must be taken into account, including target-specific capture, cost effectiveness, ease of synthesis, green chemistry, non-toxicity, biodegradability, recyclability and regeneration (the potential for recovery after use). The NPs discussed above may have some advantages, but some of them are intrinsically unstable under typical circumstances, necessitating the adoption of unique nanoscale formation methods. Agglomeration must be avoided if monodispersity is to be improved and stability increased. Another factor that may limit the use of nanomaterials is the potential toxicity of metallic NPs involved in the remediation process, as well as their byproducts and recovery costs from the remediation site. This is why, in order to create effective nanomaterial candidates that can solve environmental issues, it is crucial to have a thorough understanding of material platforms, their fabrication method and performance optimization (Guerra et al. 2018).

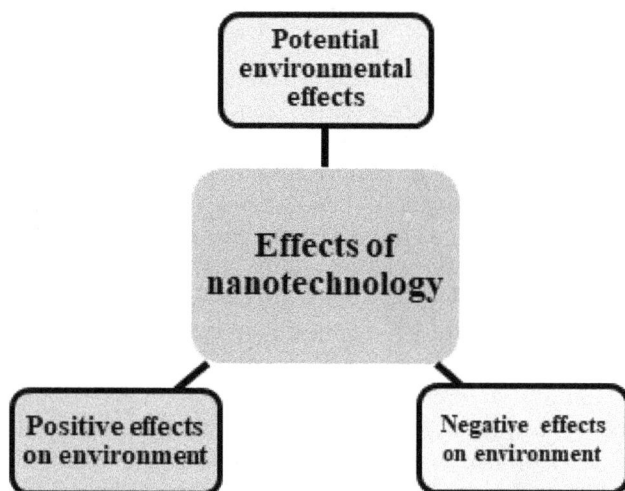

Figure 6: Effects of nanotechnology on the environment.

10. Effects of Nanotechnology on Environment

However, there are still a variety of issues that need to be resolved. Since NPs may be released and/or emitted into the environment (Figure 6), where they may persist for a long time, materials functionalized with NPs incorporated into them or deposited on their surface have a risk potential. A high level of innovation potential also exists nonetheless, for there are currently no online monitoring systems providing trustworthy real-time measurement data on the quality and quantity of NPs that are present in water in very small concentrations.

10.1 Potential Environmental Effects

Compared to bulk particles, NPs can harm the environment and the human body even more because of their larger surface areas. As a result, worry over the possible damage that NPs pose to the society has garnered attention worldwide. In addition to minimizing material use and clean-up costs, NPs are effective for customizing the properties of polymeric composite materials and for air pollution monitoring in the environment. For instance, coatings made of graphene and carbon nanotubes have been created to lessen the impacts of weathering on composite materials used in wind turbines and aircraft. Their degradation due to salt and UV exposure has been minimized by the introduction of graphene at the nanoscale. Existing materials can be given a nanoscale coating, through nanotechnology, to extend their lifespan and maintain their initial strength even in the presence of salt and upon UV exposure. Moreover, the performance of data information systems has been enhanced by the usage of carbon nanotubes.

However, there are a few potential dangers that should be taken into account when employing NPs. The nanoparticle analysis method is the main issue with nanomaterials. New and innovative nanomaterials are being developed gradually as nanotechnology advances. The shape and size of the components vary, and these are crucial aspects in determining their toxicity. The detection of NPs in air for environmental protection using the current technology is incredibly challenging due to a lack of knowledge and methodologies for characterizing nanomaterials. Toxicity of a nanomaterial can also be determined by its chemical structure and even small changes in a chemical functional group can have a significant impact on a nanomaterial's characteristics. At every level of the development of nanotechnology, a thorough risk evaluation of safety for human health and the influence on the environment is required. The exposure risk and its likelihood of exposure, toxicological analysis, transport risk, persistence risk, transformation risk and recycling potential should all be considered in the risk assessment. Another element that can be utilized to forecast the effects on the environment is the life cycle risk assessment. A product based on nanotechnology, with good experimental planning, can reduce material waste (Calvin 2003).

NPs can interact with the environment in numerous ways: It might be carried by a carrier and bio uptake, pollutants or organic substances in subsurface water. The potential of NPs for aggregation makes traditional delivery to sensitive areas possible, where these NPs can disintegrate into colloidal NPs. NPs or nanomaterials can become toxic and affect the environment in four different ways.

(a) Hydrophobic and hydrophilic NPs: TiO_2 powder is being developed by nanocoating researchers as a coating inclusion that will lessen the impacts of weathering, such as the degradation of composite materials due to salt rain. Ivana Fenogliol, among others, emphasised their worry about the need to evaluate the impact of TiO2 NPs when they are released into the environment.

(b) Mobility of contaminants: NPs can enter the atmosphere through one of the two primary routes. The fundamental source of all emissions — known as the emitter — emits NPs into the atmosphere directly. Secondary particles, such as homogenous nucleation with ammonia and sulfuric acid, are however spontaneously released.

(c) Dissolution: Scientists create and develop NPs before determining their toxicity. Many of the NPs are water soluble and difficult to remove from waste if handled improperly.

(d) Disposal: If disposed of improperly, any waste product, including nanomaterials, can have an adverse effect on the environment.

10.2 Positive Effects on the Environment

Potential advantages for the economy, society and environment are provided by nanotechnology. By offering remedies for energy use, pollution and greenhouse gas emissions, nanotechnology has the potential to lessen the impact of human activity on the environment. Potential environmental advantages of nanotechnology include cleaner and more effective industrial operations.

- Enhanced capacities to identify and remove pollutants, enhancing the quality of air, water and soil.
- High precision in manufacturing, through waste reduction.
- Clean and abundant power delivered by solar cells with higher efficiencies.
- Removing pollutants and greenhouse gases from the atmosphere.
- Less demand for substantial industrial facilities.
- Repairing ecological damages.

10.3 Negative Effects on the Environment

There is relatively little consensus regarding the risks and consequences of nanotechnology on the environment. Following is a summary of the potential environmental impacts brought on by nanotechnology:

- The enormous energy consumption brought on by the need to synthesise NPs.
- The spread of hazardous, long-lasting nanomaterials that cause environmental impacts.
- Lower rates of recycling and recovery.
- Other lifecycle stage effects on the environment (still unclear at present).
- The absence of skilled engineers and employees.

Formation, Occurrence and Emission of NPs in the Environment

A deeper comprehension of their mobility, bioavailability and toxicity is necessary to evaluate the hazards associated with employing nanomaterials in industrial products and environmental applications.

Nanomaterials have the potential for exposure, along with a hazard that develops as a result of exposure in order to constitute a risk. Point sources (such as manufacturing plants, landfills or wastewater treatment facilities) as well as non-point sources (such as items containing NPs) may release NPs. In addition to deliberate releases into the environment, inadvertent releases during production or transportation are also possible. One such intentional release involves the direct injection of chemicals into groundwater that has been contaminated with chlorinated solvents. Irrespective of whether the particles are released into soil or water directly, indirectly, through the management of waste at sewage treatment plants or through aerial deposition, they all eventually end up there. The development of aggregates and the subsequent accumulation or removal of bigger particles through sedimentation have an impact on the concentration of free NPs in the environment. As a result of exposure to NPs in the air, soil or water, or through the consumption of NP-affected plants or animals, humans can be affected by NPs directly or indirectly. Although aggregated or adsorbed NPs are less mobile, they may still be ingested by animals that live in silt or are filter feeders (Rajpoot 2021).

References

Aithal, P.S. and Aithal, S. 2015. Ideal technology concept and its realization opportunity using nanotechnology. IJAIEM 4(2): 153–164.

Aithal, P.S. and Aithal, S. 2016. Business strategy for nanotechnology based products and services. IJMSBR 5(4): 139–149.

Aldewachi, H., Chalati, T., Woodroofe, M.N., Bricklebank, N., Sharrack B. and Gardiner, P. 2018. Gold nanoparticle-based colorimetric biosensors. Nanoscale 10: 18–33.

Arora, A., Choudhary, V. and Sharma, D.K. 2011. Effect of clay content and clay/surfactant on the mechanical, thermal and barrier properties of polystyrene/organoclay nanocomposites. J. Polym. Res. 18: 843–857.

Azeredo, H.M.C., Mattoso, L.H.C., Wood, D., Williams, T.G., Avena-Bustillos R.J. and McHugh, T.H. 2009. Nanocomposite edible films from mango puree reinforced with cellulose nanofibers. J. Food Sci. 74: N31–N35.

Bajpai, V.K., Kamle, M., Shukla, S., Mahato, D.K., Chandra, P., Hwang, S.K. et al. 2018. Prospects of using nanotechnology for food preservation, safety, and security. J. Food Drug Anal. 26.

Bawa, R., Audette, G.F. and Rubinstein, I. 2016. Handbook of clinical nanomedicine: nanoparticles, imaging, therapy, and clinical applications. Pan Stanford.

Bhattacharya, S., Jang, J., Yang, L., Akin, D. and Bashir, R. 2007. Biomems and nanotechnology-based approaches for rapid detection of biological entities. J. Rapid Methods Auto Microb. 15: 1–32.

Brody, A.L. 2007. Case studies on nanotechnologies for food packaging. Food Technol. 61: 102–107.

Bucking, M., Hengse, A., Gruger, H. and Schulte, H. 2017. Smart systems for food quality and safety. *In*: Monique, A.V.A. and de Voorde, M.V. (eds.). Nanotechnology in Agriculture and Food Science. WileyVCH, Oxford.

Byrne, J.D. and Baugh, J.A. 2008. The significance of nanoparticleinduced pulmonary fibrosis. Mcgill. J. Med. 11(1): 4350.

Bystrzejewska-Piotrowska, G., Golimowski, J. and Urban, P.L. 2009. Nanoparticles: Their potential toxicity, waste and environmental management. J. Waste Manag. 29(9): 2587–2595.

Calvin, V.l. 2003. The potential environmental impact of engineered nanomaterials. Nat. Biotecnol. 21: 1166–1170.

Campbell, M.L., Guerra, F.D., Dhulekar, J., Alexis F. and Whitehead, D.C. 2015. Target-specific capture of environmentally relevant gaseous aldehydes and carboxylic acids with functional nanoparticles. Chem. A Eur. J. 21: 14834–14842.

Chaudhry, Q., Scotter, M., Blackburn, J., Ross, B., Boxall, A., Castle, L. et al. 2008. Applications and implications of nanotechnologies for the food sector. Food Addit. Contam. Part A Chem. Anal. 25(3): 241–258.

Cheng, C., Chen, H.Y., Wu, C.S., Meena, J.S., Simon T. and Ko, F.H. 2016. A highly sensitive and selective cyanide detection using a gold nanoparticle-based dual fluorescence–colorimetric sensor with a wide concentration range. Sensors and Actuators B: Chemical 227: 283–290.

Cushen, M., Kerry, J., Morris, M., Cruz-Romero M. and Cummins, E. 2012. Nanotechnologies in the food industry—recent developments, risk and regulation. Trends Food Sci. Technol. 24: 30–46.

Cushen, M., Kerry, J., Morris, M., Cruz-Romero, M. and Cummins, E. 2012. Nanotechnologies in the food industry—recent developments, risk and regulation. Trends Food Sci. Technol. 24: 30–46

Dasgupta, N., Ranjan, S., Patra, D., Srivastava, P., Kumar A. and Ramalingam, C. 2016. Bovine serum albumin interacts with silver na- noparticles with a "side-on" or "end on" conformation. Chem. Biol. Interact. 253: 100–111.

Deng, D., Jin, Y., Chen, Y., Qi, T. and Xiao, F. 2012. Preparation of copper nanoparticles with low sintering temperature. Paper Presented at 14th International Conference on Electronic Materials and Packaging (EMAP), pp. 1–4.

Dhawan, A., Sharma, V. and Parmar, D. 2009. Nanomaterials: a challenge for toxicologists. Nanotoxicology 3(1): 1–9.

DiSia, P. 2017. Nanotechnology Among Innovation, Health and Risks. Procedia Soc. Beh. Sci. 237: 1076–1080.

Dobrucka, R. 2014. Application of nanotechnology in food packaging. J. Microbiol. Biotechnol. Food Sci. 3: 353–359.

Feng, C., Khulbe, K.C., Matsuura, T., Tabe, S. and Ismail, A.F. 2013. Preparation and characterization of electro-spun nanofiber membranes and their possible applications in water treatment. Sep. Purif. Technol. 102: 118–135

Fenoglio, I., Greco, G., Livraghi, S. and Fubini, B. 2009. Non-UV-induced radical reactions at the surface of TiO2 nanoparticles that may trigger toxic responses. Chemistry–A European Journal 15(18): 4614–4621.

Freitas, R.A. 2005. What is Nanomedicine? Nanomedicine: Nanotech. Biol. Med. 1(1): 29.

Gabr, M.H., Okumura, W., Ueda, H., Kuriyama, W., Uzawa, K. and Kimpara, I. 2015. Mechanical and thermal properties of carbon fiber/polypropylene composite filled with nano-clay. Compos. Part B 69: 94–100.

Gardner, J. and Dhai, A. 2014. Nanotechnology and water: Ethical and regulatory considerations. *In*: Ajay Kumar Mishra (ed.). Application of Nanotechnology in Water Research. Massachusetts. Scrivener Publishing LLC. 120.

Gardner, J. 2015. Nanotechnology in medicine and healthcare: Possibilities, progress and problems: education and training. S. Afr. J. Bioeth. Law. 8(2): 50–53.

Grassian, V.H., Shaughnessy, P.T., Adamcakova-Dodd, A., Pettibone J.M. and Thorne, P.S. 2007. Inhalation exposure study of titanium dioxide nanoparticles with a primary particle size of 2 to 5 nm. Environ. Health Perspect. 115(3): 397–402.

Guerra, F.D., Attia, M.F., Whitehead D.C. and Alexis, F. 2018. Nanotechnology for environmental remediation: Materials and applications. Molecules 23(7): 1760.

Guerra, F.D., Smith, G.D., Alexis, F. and Whitehead, D.C. 2017. A survey of VOC emissions from rendering plants. Aerosol Air Qual. Res. 17: 209–217.

Hall, R.H. 2002. Biosensor technologies for detecting microbiological food borne hazards. Microbes Infect. 4: 425–432.

Hallock, M.F., Greenley, P., DiBerardinis, L. and Kallin, D. 2009. Potential risks of nanomaterials and how to safely handle materials of uncertain toxicity. J. Chem. Health Saf. 16(1): 16–23.

He, X., Deng, H. and Hwang, H. 2019. The current application of nanotechnology in food and agriculture. J. Food Drug Anal. 27: 1–21.

Hulla, J.E., Sahu, S.C. and Hayes, A.W. 2015. Nanotechnology: History and future. Hum. Exp. Toxicol. 34(12): 1318–1321.

Hutchison, J.E. 2016. The road to sustainable nanotechnology: Challenges, progress and opportunities. ACS Sustain. Chem. Eng. 4: 5907–5914.

Jagadevan, S., Jayamurthy, M., Dobson, P. and Thompson, I.P.A. 2012. Novel hybrid nano zerovalent iron initiated oxidation – biological degradation approach for remediation of recalcitrant waste metalworking fluids. Water Research 46: 2395–2404.

Jahed, F.S. and Hamidi, S. 2020. Applications of surface plasmon resonance in human health care. Nanomedicine 15(19): 1823–1827.

Jain, K.K. 2008. The Handbook of Nanomedicine. Springer, New York.

Jain, K.K. 2004. Applications of biochips: From diagnostics to personalized medicine. Curr. Opin. Drug Discov. Dev. 7(3): 285–289.

Joye, I.J., Davidov-Pardo, G. and McClements, D.J. 2016. Nanotechnology in food processing. pp. 49–55. *In*: Caballero, B., Finglas, P.M. and Toldrá, F. (eds.). Encyclopedia of Food and Health. Academic Press: Oxford, UK. ISBN 978-0-12-384953-3.

Kim, S.W. and Cha, S.H. 2014. Thermal, mechanical, and gas barrier properties of ethylene—Vinyl alcohol copolymer-based nanocomposites for food packaging films: Effects of nanoclay loading. J. Appl. Polym. Sci. 131.

Kootenaei, F.G., Aminirad, H., Ramezani, M. and Mehrdadi, N. 2013. Applications of Nanotechnology in Environmental Engineering. Global Journal on Advances Pure and Applied Sciences 1.

Kumar, C.S.S.R. 2006. Nanomaterials for Biosensors. Wiley-VCH, Weinheim.

Kuswandi, B. 2017. Environmental friendly food nano-packaging. Environ. Chem. Lett. 15: 205–221.

Lau, H.C., Yu, M. and Nguyen, Q.P. 2017. Nanotechnology for oilfield applications: Challenges and impact. J. Pet. Sci. Eng. 157: 1160–1169.

Li, C., Zhou, K., Qin, W., Tian, C., Qi, M., Yan, X. et al. 2019. A review on heavy metals contamination in soil: Effects, sources, and remediation techniques. Soil and Sediment Contamination: An International Journal 28(4): 380–394.

Li, H., Xu, C. and Banerjee, K. 2010. Carbon nanomaterials: The ideal interconnect technology for next-generation ICs. IEEE Design & Test of Computers 27(4): 20–31.

Liu, X., Vinson, D., Abt, D., Hurt, R.H. and Rand, D.M. 2009. Differential toxicity of carbon nanomaterials in drosophila: Larval dietary uptake is benign, but adult exposure causes locomotor impairment and mortality. Environ. Sci. Techno. 43(16): 6357–6363.

Liu, B., Zhuang, J. and Wei, G. 2020. Recent advances in the design of colorimetric sensors for environmental monitoring. Environmental Science: Nano. 7(8): 2195–2213.

Llorens, A., Lloret, E., Picouet, P.A., Trbojevich, R. and Fernandez, A. 2012. Metallic based micro and nanocomposites in food contact materials and active food packaging. Trends Food Sci. Technol. 24: 19–29.

Loiseau, A., Zhang, L., Hu, D., Salmain, M., Mazouzi, Y., Flack R. et al. 2019. Core–shell gold/silver nanoparticles for localized surface plasmon resonance-based naked-eye toxin biosensing. ACS Appl. Mater. 11(50): 46462–46471.

Mabeck, J.T. and Malliaras, G.G. 2006. Chemical and biological sensors based on organic thin-film transistors. Anal. Bioanal. Chem. 384(2): 343–353.

Mamo, T., Moseman, E.A., Kolishetti, N., Salvador-Morales, C., Shi, J., Kuritzkes, D.R. et al. 2010. Emerging nanotechnology approaches for HIV/AIDS treatment and prevention. Nanomedicine 5(2): 269285.

Mansoori, G.A., Bastami, T.R., Ahmadpour, A. and Eshaghi, Z. 2008. Environmental application of nanotechnology. Annual Review of Nano Research, 439–493.

Masciangoli, T. and Zhang, W. 2003. Environmental technologies. Environ. Sci. Technol. 37: 102–108.

Mathuria, J.P. 2009. Nanoparticles in tuberculosis diagnosis, treatment and prevention: A hope for the future. Dig. J. Nanomater Biostruct. 4(2): 309312.

Mbunge, E., Muchemwa, B. and Batani, J. 2021. Sensors and healthcare 5.0: Transformative shift in virtual care through emerging digital health technologies. J. Glob. Health. 5(4): 169–177.

Michelson, E.S., Sandler, R. and Rejeski, D. 2008. Nanotechnology. *In*: Crowley, M. (ed.). From Birth to Death and Bench to Clinic: The Hastings Center Bioethics Briefing Book for Journalists, Policy Makers, and Campaigns. Garrison, NY: The Hastings Center. 111116.

Mohanty, S.P. and Kougianos, E. 2006. Biosensosrs: A tutorial review. IEEE Potentials 25(2): 35–40.

Musee, N., Brent, A.C. and Ashton, P.J. 2010. A South African research agenda to investigate the potential environmental health and safety risks of nanotechnology. S. Afr. J. Sci. 106: 34.

Nagar, A. and Pradeep, T. 2020. Clean water through nanotechnology: Needs, gaps, and fulfillment. ACS Nano 14(6): 6420–6435.

Nile, S.H., Baskar, V., Selvaraj, D., Nile, A., Xiao J. and Kai, G. 2020. Nanotechnologies in food science: Applications, recent trends, and future perspectives. Nano Micro Lett. 12: 45.

Pacheco-Torgal, F. and Jalali, S. 2011. Nanotechnology: Advantages and drawbacks in the field of construction and building materials. Constr. Build. Mater. 25(2): 582–590.

Paisoonsin, S., Pornsunthorntawee O. and Rujiravanit, R. 2013. Preparation and characterization of ZnO-deposited DBD plasma-treated PP packaging film with antibacterial activities. Appl. Surf. Sci. 273: 824–835.

Pandey, B. and Fulekar, M.H. 2012. Nanotechnology: Remediation Technologies to clean up the environmental pollutants. Res. J. Chem. Sci. 2: 90–96.

Qu, X., Alvarez, P.J. and Li, Q. 2013. Applications of nanotechnology in water and wastewater treatment. Water Research 47: 3931–3946.

Rabea, E.I., Badawy, M.E., Stevens, C.V., Smagghe, G. and Steurbaut, W. 2003. Chitosan as antimicrobial agent: Applications and mode of action. Biomacromolecules 4(6): 1457–1465.

Raj, P., Sharma, H., Mishra, D., Murali, K.P., Han, K., Swaminathan, M. et al. 2012. Nanomagnetics for high-performance, miniaturized power, and RF components [nanopackaging]. IEEE Nanotechnol. Mag. 6(3): 18–23

Rajpoot, S. 2021. Impact of nanotechnology on environment – A review. Sci. Technol. 7: 159–164.

Ramirez-Frometa, N. 2006. Cantilever biosensors. Biotechnol. Apl. 23: 320–323.

Ranjan, S., Dasgupta, N., Chakraborty, A.R., Melvin Samuel, S., Ramalingam, C., Shanker, R. et al. 2014. Nanoscience and nanotechnologies in food industries: Opportunities and research trends. J. Nanopart. Res. 16: 2464.

Rhim, J.W., Park, H.M. and Ha, C.S. 2013. Bio-nanocomposites for food packaging applications. Prog. Polym. Sci. 38(10): 1629–1652

Sharma, V. and Sharma, A. 2012. Nanotechnology: An emerging future trend in wastewater treatment with its innovative products and processes. International Journal of Enhanced Research in Science Technology and Engineering 1: 121–128.

Singh, N., Manshian, B., Jenkins, G.J., Griffiths, S.M., Williams, P.M., Maffeis T.G. et al. 2009. NanoGenotoxicology: The DNA damaging potential of engineered nanomaterials. Biomaterials 30(23-24): 3891–3914.

Singh, T., Shukla, S., Kumar, P., Wahla, V., Bajpai, V.K. and Rather, I.A. 2017 Application of nanotechnology in food science: Perception and overview. Front. Microbiol. 8. doi:10.3389/fmicb.2017.01501.

Sorrentino, A., Gorrasi, G. and Vittoria, V. 2007. Potential perspectives of bionanocomposites for food packaging applications. Trends Food Sci. Technol. 18: 84–95.

Sozer, N. and Kokini, J.L. 2009. Nanotechnology and its applications in the food industry. Trends Biotechnol. 27: 82–89.

Thomas, V., Yallapu, M.M., Sreedhar, B. and Bajpai, S.K. 2007. A versatile strategy to fabricate hydrogelsilver nanocomposites and investigation of their antimicrobial activity. J. Colloid. Interface Sci. 315(1): 389–395

Tratnyek, P.G. and Johnson, R.L. 2006. Nanotechnologies for environmental cleanup. Nano Today 1: 44–48.

Tyagi, M. and Tyagi, D. 2014. Polymer nanocomposites and their applications in electronics industry. Int. J. Electron. Electr. Eng. 7(6): 603–608.

Tyshenko, M.G. and Krewski, D. 2008. A risk management framework for the regulation of nanomaterials. Int. J. Nanotechnol. 5(1): 143–160.

Vaseashta, A., Vaclavikova, M., Vaseashta, S., Gallios, G., Roy, P. and Pummakarnchana, O. 2007. Nanostructures in environmental pollution detection, monitoring, and remediation. Sci. Technol. Adv. Mater. 8: 47–59.

Wegmann, M., Michen, B. and Graule, T. 2008. Nanostructured surface modification of microporous ceramics for efficient virus filtration. J. Eur. Ceram. Soc. 28: 1603–1612.

Yashwant, Y.S., Deepika, D.C. and Tansukh, T.B. 2022. Impact of nanotechnology on environment and their role in agronomy and food stuffs production: An Overview: Role of nanotechnology in agronomy. Int. J. Biomater. 1(2): 1–4.

Yen, C.W., dePuig, H., Tam, J.O., Gómez-Márquez, J., Bosch, I., Hamad-Schifferli, K. et al. 2015. Multicolored silver nanoparticles for multiplexed disease diagnostics: Distinguishing dengue, yellow fever, and Ebola viruses. Lab on a Chip. 15(7): 16381641.

Index

For Product Safety Concerns and Information please contact our EU
representative GPSR@taylorandfrancis.com
Taylor & Francis Verlag GmbH, Kaufingerstraße 24, 80331 München, Germany